Mr. Bloomfield's Orchard

Mr. Bloomfield's Orchard

The Mysterious World of Mushrooms, Molds, and Mycologists

NICHOLAS P. MONEY

2002

OXFORD
UNIVERSITY PRESS

Oxford New York
Auckland Bangkok Buenos Aires Cape Town Chennai
Dar es Salaam Delhi Hong Kong Istanbul Karachi Kolkata Kuala Lumpur
Madrid Melbourne Mexico City Mumbai Nairobi São Paulo
Shanghai Singapore Taipei Tokyo Toronto

and an associated company in Berlin

Published by Oxford University Press, Inc.
198 Madison Avenue, New York, New York 10016

www.oup.com

Library of Congress Cataloging-in-Publication Data
Money, Nicholas P.
Mr. Bloomfield's orchard : a personal view
of fungal biology / Nicholas P. Money.
p. cm. Includes bibliographical references.
ISBN 0-19-515457-6
1. Fungi. I. Title.
QK603 .M59 2002
579.5—dc21 2002072654

1 3 5 7 9 8 6 4 2
Printed in the United States of America
on acid free paper

For Terence Ingold and his jewels

Contents

Preface

It is indeed a singular and despised family to the history of which we are about to dedicate this volume.

—M. C. Cooke, *British Fungi* (1871)

Some time ago, my colleague Jerry McClure told me that the most fortunate among us are faced with three options at the juncture in life once valued as the midlife crisis: go insane, engage in an extramarital affair, or write a book. In my own approach to this disconcerting landmark, all but the third option vaporized under my wife's guidance. The fruit of her influence is in your hands.

Mr. Bloomfield's Orchard is a personal reflection on the subject of mycology, the scientific study of fungi. Many people giggle at the mention of these organisms, drawing on vague notions about hallucinogens and poisons, fairy tales, and the erectile behavior of mushrooms. Although such peculiarities may draw people to this book, my primary concern as its author is to explore our profound intimacy with fungi and to articulate the most important consequences of these interactions. Employing a flexible interpretation of that term *interaction*, this is a celebration both of the fungi (even the nasty ones) and of a selection of the scientists obsessed with their study (none that I know of have been exceptionally nasty). While I have written for a general audience, particularly those with some scientific education, I also hope to deepen the appreciation of fungi among my biologist peers.

There are a number of people to whom I extend deep gratitude for stimulating this book. As a teenager studying at Bristol University, my first—and most inspiring—guide to mycology was Mike Madelin, and my admiration for my doctoral mentor at Exeter, John Webster, grows with every year. The dedication of this book to Terence Ingold is explained in the narrative. I also thank the staff of the Lloyd Library in Cincinnati for maintaining the world's supreme archive of mycological publications. This book would not have been possible without the sanctuary offered by the Lloyd. Speaking of sanctuaries, Frank Harold was kind enough to offer me one in his laboratory in Colorado at a time when I was lost in New England, and has now shown great generosity in reviewing the *Bloomfield* manuscript. I thank my wife, Diana Davis, for agreeing to marry me, and more pertinently in the context of this book, for her invaluable service as my primary reader.

By discussing fungal processes that I have investigated (if only peripherally), this book has enabled me to revisit my twenty-year journey from student to professional mycologist. I hope you have as much fun reading about this odyssey as I have had recreating it.

Nicholas P. Money
Oxford, Ohio
January 2002

Mr. Bloomfield's Orchard

CHAPTER 1

Offensive Phalli and Frigid Caps

I am . . . a mushroom
On whom the dew of heaven drops now and then.
—John Ford, *The Broken Heart* (1633)

All sound in the forest is damped by a morning mist trapped under the pine trees on the edge of the moors in Devon, England. Three men are tramping up a steep slope, their boots sinking into the soaking needles. They are searching for eggs. A dead deer smell hangs in the watery air, a hint of sweetness too, and even a suggestion of semen. This odor cannot be ignored. Steamed glasses are wiped every few minutes. The oldest of the men is wearing hunting pants that end at the knees, thick hiking socks bridging the gaps to his red-laced boots. Webster stops, his blue eyes bulging as he scans the forest floor. Squatting, he parts the pine needles and uncovers five pure white eggs, somewhat larger than golf balls. Each is attached to the soil by a branched umbilical cord that snaps as it is tugged away from its siblings. The jelly-filled spheres have cold skins. What monsters will hatch from such spawn? And what is that smell?

A few feet from the nest is a very ugly penis. Poking 6 inches or more from the pine needles, a full erection that arches a little, a pallid shaft protruding from a broken egg. Its head glistens with green-black syrup (Figure 1.1). This is the source of the smell. At the tip, a small hole is circled by a raised ring. Some degenerate must be hiding under the needles and is evidently aroused by the experience. But wait a moment; there are hundreds of these apparitions higher up the slope. Have the collectors wandered into a colony of sexual deviants fixated upon live burial?

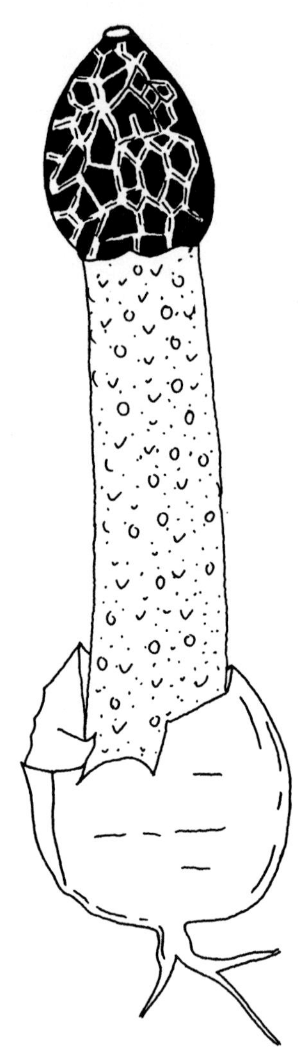

Fig. 1.1 Erect fruiting body of *Phallus impudicus*.

But there are no horny corpses. Little Red Riding Hood's chastity is safe. The erections were accomplished by a fungus whose Latin name is *Phallus impudicus*, the shameless penis, a type of "stinkhorn." You must not forego the spectacle offered by this beast. My first encounter with this bizarre species was made during a foray with the mycologist John Webster (Figure 1.2) and his Spanish assistant Henry Descals. The site on Dartmoor was a favorite of John's, a place he visited every year to collect specimens for his undergraduate classes at the University of Exeter.

Fig. 1.2 John Webster.

Phallic mushrooms belong to the large group of fungi that includes the more familiar organisms that generate brackets on trees and buttons and portabella caps that end their lives sautéed in olive oil. These organisms are members of a group of fungi called the Basidiomycota,[1] a name that refers to a special kind of spore or microscopic seed called the basidiospore. Thirty thousand species of basidiomycete have been described by scientists, and seventy or so are phallic mushrooms and related fungi that manufacture smelly cages. The phallic ones have proven impossible to ignore. They are featured in Pliny the Elder's thirty-seven-volume *Natural History* written in the first century A.D., a publication with the modest goal of recording "all the contents of the entire world." In his seventeenth-century herbal, John Gerard pictured them in a modest, tip-down orientation, with the following description: "*Fungus virilis penis arecti forma,* which wee English, [call] Pricke Mushrum, taken from his forme." For Victorians in England, sufficiently obsessed with sex to become excited by table legs, their appearance was too much to bear. As a mature woman, Charles Darwin's daughter Etty so despised stinkhorns that she mounted an antifungal jihad with the aid of gloves and a pointed stick. She burned the collections in secret, thereby protecting the purity of thought among her female servants.

The transformation from egg to stinking horn is a slow erection that often begins in the cool of the night and is not complete until sunrise. If an unhatched egg is cut in half, the tissues of the expanded structure are displayed in prefabricated form (Figure 1.3). A hollow shaft of white spongy material called the receptacle runs pole-to-pole through its center. The receptacle is surrounded by the green-black cushion of spores called the gleba, cased in a clear jelly veiled with white skin. When the egg hatches, the receptacle expands by absorbing water and ruptures the skin, carrying the spores on its tip into the air. The jelly lubricates the extending shaft and helps keep the mass of spores in place. The spores are embedded in slime that contains a cocktail of volatile chemicals, including hydrogen sulfide, formaldehyde, methylmercaptan, and unique compounds called phallic acids. Impersonating the smell of rotting flesh, the stinkhorn is irresistible to flies, which swarm on the head, and to slugs, which glide for 20 or more feet for the reward of the cadaverous confection. Within a few hours, the head is cleaned down to the dimpled white surface of receptacle tissue, and the shaft begins to wilt. Although the marathon erection is over, the stinkhorn has been successful. Flies and slugs carry and defecate its spores, whose stinkhorn genes contain the information needed to make more stinkhorns. In common with humans, stinkhorns are here because they are very good at making copies of themselves.

Stinkhorns and other mushrooms are the tips of mycological icebergs. The umbilical cord at the bottom of the egg connects with the larger organism that pulses unseen through leaf litter, crawls under the bark of dying trees, and connects with the roots of healthier ones. This is the feeding phase of the organism's life, or life cycle, and grows as masses of filamentous cells called hyphae. Only when these hyphae have gathered a sufficient harvest of food, and when the subterranean fungus is fattened and pumped full of water, can it surface to disturb our composure.

Biologists decipher the shape and structure of different organisms by thinking about the functions for which they may be adapted, or the challenges that have been overcome by developing in a particular way. The apparently ornamental figure of the phallic mushroom is really a very conservative structure. The top of the shaft is a sensible location for the spore mass because its pungent slime is concentrated where it acts best as a beacon to flies. Stinkhorn receptacles are very delicate structures. They are built

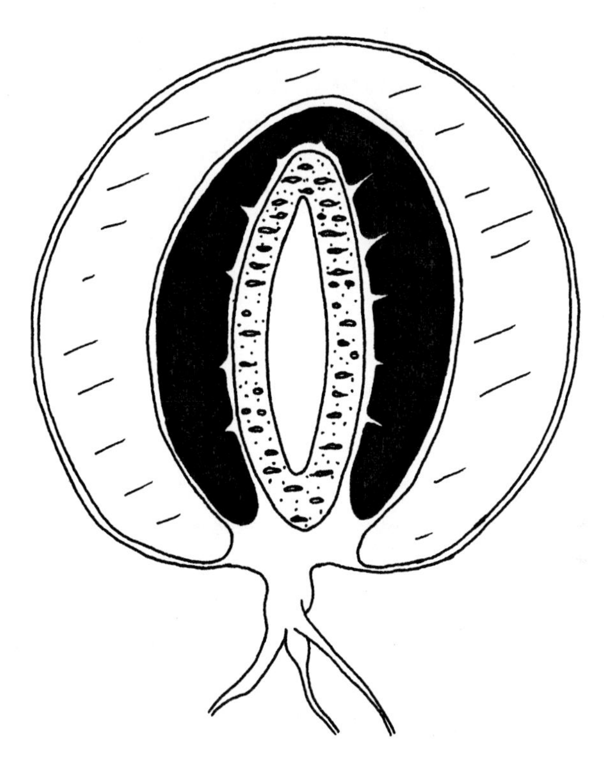

Fig. 1.3 Cut egg of phallic mushroom. The central receptacle, which expands to form the stalk, is surrounded by the green-black mass of developing spores called the gleba. Jelly surrounds the gleba.

from masses of corrugated hyphae that are stretched into a weft of filaments when the egg hatches. Most of the volume of the erect fruiting body is air. But mechanically speaking, the stinkhorn is comparable with the mammalian penis because both erections are maintained by pressurized fluid rather than a column of solid tissue. The penis contains flattened reservoirs that become engorged with blood, while the tissue of the stinkhorn receptacle is built to tear apart to make a honeycomb supported by pressurized water within its hyphae. Despite these similarities, the origin of the pressurized fluid is fundamentally different in the two structures. Penile blood pressure is generated by muscular activity; stinkhorn pressure is osmotic in origin, something akin to the way that water is soaked up into a dry sponge. While we can deconstruct the stinkhorn and explain its parts, the extraordinary phallic resemblance remains a great surprise. I suppose that this unusual fruiting body may be a jest by Satan—in its various stages of development, *Phallus* has been identified as the devil's eggs, devil's horn, and devil's stinkpot[2]—but I'm putting my money on the Darwinian explanation. At least for the fungus, fruiting bodies function to produce and disperse spores, nothing else.

Mycologists have described thirty truly phallic-looking mushroom species. As its common name suggests, the dog stinkhorn, *Mutinus caninus*, is smaller in stature, has a pink shaft, and lacks the bulbous head. It still smells awful and attracts flies. Species of *Dictyophora* are recognized by a lacy veil that hangs down as a skirt beneath the head (see jacket photo). The crinoline feminizes the phallic effect a little, and may offer a ladder that allows wingless insects to reach the spores by crawling from surrounding plants. The eggs of one species of *Dictyophora* are sold as delicacies in China and are also marketed as aphrodisiacs. Inside the egg, stinkhorn slime does not smell too awful, and some authors of mushroom guidebooks claim that the whole thing can be consumed without much suffering. In his book, *In the Company of Mushrooms*, Elio Schaechter[3] admitted to enjoying stinkhorn eggs and remarked that once filled with cream, rings cut from the expanded receptacles were delicious. On a more general note, it is a tragedy in a country as populous as China that anything from tiger turds to whale afterbirths can be sold as long as the suggestion is made that their consumption enhances erectile function.

The related cage fungi produce other kinds of flamboyant fruiting bodies that share the seductive power that phallic mushrooms wield over insects. Again, a preformed receptacle is packaged into an egg, and as this structure absorbs water and expands, it carries a stinking spore mass into the air. Rather than exiting the egg as a single shaft, the receptacles of cage fungi unfold into more open structures. *Clathrus* forms a spherical cage with spores spread on the inside of its bars (Figure 1.4 a). The receptacle of *Anthurus* separates into four or more arms that curl back over the egg to create a star (Figure 1.4 b). The arms are bright orange and their inner surface is smeared with the spores. A time-lapse video that shows the hatching of an *Anthurus* egg is quite shocking. It is difficult to describe the performance delivered by this fungus. There is nothing comparable. Here's my best shot: as this fruiting body issues from the ground, its livid arms simulate the agonized contortions of a horribly injured lobster. Other cage fungi form stalks with chambered heads or claws at their summit, and *Laternea* elaborates long arms like *Anthurus* but fuses them at their tips and dangles a reeking lantern inside the resulting vault (Figure 1.4 c). Flies are the usual vectors for spore dispersal, but ants and stingless bees have also been seen feeding on some cages.

Ileodictyon (intestinal net) is a cage fungus that grows in New Zealand and Australia. The Maori were quite taken with this fruiting body, according it nine different names and barbecuing its eggs. When it escapes human consumption, the white *Ileodictyon* cage expands from a buried egg and disengages from its papery skin. The detached cage, smeared with the usual excremental spore gunk, is then blown about on the surrounding grass. The Maori didn't eat the repugnant hatchlings, denigrating them as the "feces of ghosts or of the stars." The quote is taken from an intriguing article by the distinguished British mycologist Graham Gooday (a delightful scientist whose appearance evokes the stereotypical image of a Royal Air Force fighter pilot from the Second World War) and his friend John Zerning.[4] Zerning was struck by the shape of a dried specimen of *Ileodictyon* displayed by Gooday at a meeting. He noticed the resemblance between the cage and the geodesic homes designed by Richard Buckminster Fuller (Bucky), which became popular in the 1960s. This polyhedral form is also characteristic of the carbon-based molecules called buckminsterfullerenes, or buckyballs, which come close to making organic chemistry seem interesting. The similarity of hippy dwellings, buckyballs, and ghost feces is a reflection of the surprising strength offered by their lightweight polyhedral structure. Any weight saving is valuable for a fungus that, by necessity, makes conservative use of building materials. The resistance to compression of the *Ileodictyon* cage is important during its emergence from buried eggs and also when it is blown around. By maintaining an open shape, the receptacle provides a large surface area for exposure of the spores and their fetid scent.

Small changes in the details of receptacle development probably account for the great variety of mature fruiting body shapes in stinkhorns and cage fungi. For example, weakening of tissue along four or five tracks running the length of the receptacle would cause the shaft to split like a banana skin into four or five arms upon pressurized expansion. This would require alterations in the arrangement of the receptacle tissues inside the egg, or changes in the activity of specific enzymes during hatching. Then, with the mobilization of some genes to control the orange coloration of the receptacle, a *Phallus*-type fruiting body would be transformed into *Anthurus*.

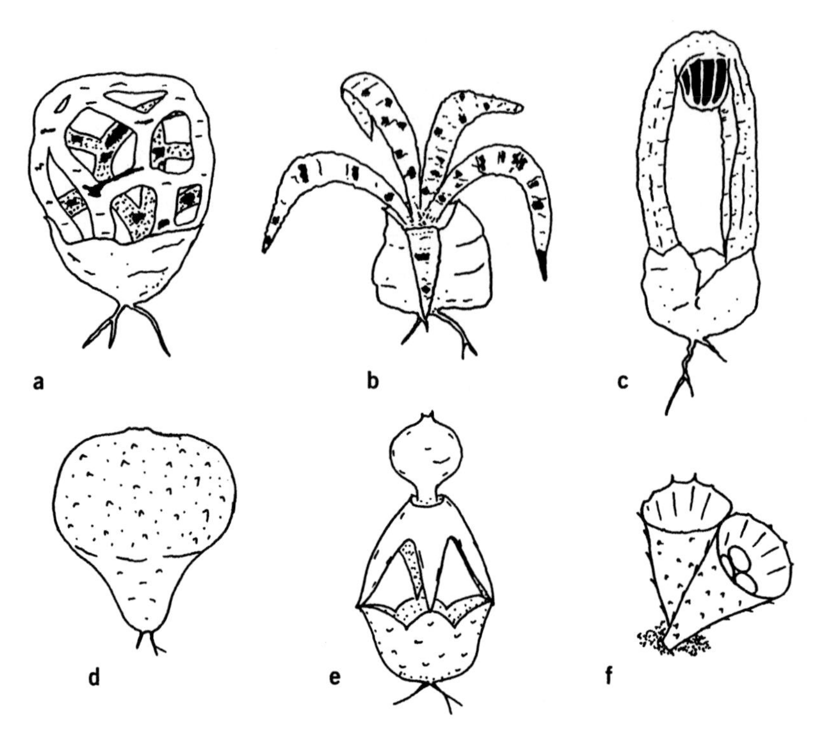

Fig. 1.4 Fruiting bodies of various gasteromycete fungi. (a–c) cage fungi: (a) *Clathrus ruber*; (b) *Anthurus archeri*; (c) *Laternea triscapa*; (d) puffball, *Lycoperdon perlatum*; (e) earth-star, *Geastrum fornicatum*; (f) bird's nest fungus, *Cyathus striatus*. Not drawn to the same scale.

This is an oversimplification of the developmental differences between these organisms, because there are other microscopic distinctions between their structures. But research on other fungi does suggest that conspicuous modifications in fruiting body morphology can be derived by surprisingly minor changes in the expression of enzymes during development.

Given the similarities among all of the phallic and cage fungi, it seems probable that natural selection may have sculpted the existing species in a relatively short period of time, perhaps in as little as a few million years. But why did all these structures evolve? Why did a phallus that divides at its tip evolve from an ancestor that did not, or vice versa? The answer surely lies in the relationships between these fungi and the insects and other invertebrates that disperse their spores. Different species of flies are lured by particular scents and personalized visual cues, so the various receptacles probably reflect distinctive solutions to the challenge of supporting and advertising spore slime. Biologists already recognize the significance of analogous characteristics in the origins of flowers among insect-pollinated plants. While humans are seduced by many floral perfumes, colors, and shapes, there are also numerous insect-pollinated flowers, such as the Sumatran giant *Amorphophallus titanum*, or corpse flower, which emit stinkhorny smells.[5] Stinkhorns, cage fungi, and putrid flowers have all evolved parallel features that attract insects that ordinarily congregate around carrion.

Along with the stinkhorns and cage fungi, other organisms including puffballs, earth-balls, earth-stars, and bird's nest fungi belong to the gasteromycete section of the basidiospore-producing fungi (Figure 1.4 d–f). Surpassing the inventions of all other fungi, the gasteromycetes have evolved a circus of mechanisms for dispersing their spores. Adapting an image from Richard Dawkins, baby stinkhorns use insect wings to fly away from their parents.[6] The offspring of puffballs, earth-balls, and earth-stars are puffed into the air and are then carried away by wind. Bird's nest fungi also use a two-stage dispersal mechanism. Their tiny fruiting bodies are shaped like champagne flutes and contain packets of spores called peridioles. Raindrops splash the peridioles from these cups onto surrounding blades of grass. Unsuccessful spores, those destined for a swift passage to stinkhorn heaven or hell, wait, and wait longer, and dehydrate, and die. Fortunate ones are consumed by herbivores grazing around the fruiting bodies, are carried by the animals as they pass through their digestive systems, and later deposited in a convenient pat of warm manure. Cow feces offer perfect residence for a young bird's nest fungus (suggesting that stinkhorn hell lacks the excrement-filled ditch of Dante's *Inferno*). Finally, one gasteromycete fungus shoots a black ball of spores from a fruiting body that operates as a tiny trampoline. This organism, called *Sphaerobolus*, grows on wood mulch, and can ruin the paintwork of a car parked close to a wet flower bed. The spore balls stick to smooth surfaces with incredible tenacity, and even when they are removed by vigorous cleaning, spots remain in the paint. Like the bird's nest fungi, this villain is adapted for an excursion through a herbivore gut, but it doesn't help to know that the intended targets of *Sphaerobolus* are grass blades rather than my beloved Ford Probe.

The gasteromycetes are defined by the fact that their spores form inside the fruiting body rather than on gills or other fertile surfaces exposed to the air. Their scientific name refers to this developmental feature: gastero = stomach, mycetes = fungi, stomach fungi. They seem to have evolved from different kinds of ancient fungi that produced conventional umbrella-shaped mushrooms, and as such are regarded as a ragbag of species rather than a natural grouping of organisms. The natural group is an important concept in biology. Contrary to the delusions of Christian fundamentalists, all animals with nipples and fur, for example, are descendants of a single ancestral species. They belong to a natural group, the mammals, from which every other living thing is excluded: without nipples you don't even merit an interview. Time is a crucial consideration in this discussion, because, of course, every pair of species shares an ancestor that could be found by delving back far enough into their respective evolutionary histories. Humans and stinkhorns are certainly related, and far more closely (according to their genes) than either is to any plant. But the natural group that includes *Homo* and *Phallus* also encompasses every animal and every fungus, and as such is a pretty esoteric gathering.

In common with stinkhorns, gilled mushrooms are devices for spore production and dispersal, nothing more or less. They have always held great fascination for me, and I suppose my deepest professional roots lie in childhood tales involving mushrooms. My earliest memory of a mycological experience comes from a dentist's office. I was 5 years old and under gas for multiple tooth extractions when I hallucinated a fairy ring with elves and other phantasms dancing around the mushrooms. Then I awoke, tumbling down the stairs from the torture chamber, bloody handkerchief pressed to my mouth. I have remained captivated by the eeriness of mushrooms, and have joined the ranks of mycologists who have become fascinated by trying to understand how they operate. This is not a simple matter.

Umbrella- and bracket-shaped mushrooms maximize their spore-producing capacity for a minimal investment in fruiting body tissue by supporting massive numbers of spores with a single stalk. These fungi spread their fertile tissues underneath the cap, folding a vast spore-producing mat called the hymenium over the surface of gills, ripples, or spines, or inside tubes. If a thin slice is cut from a mushroom cap with a razor blade and viewed under a microscope, spore-producing cells called basidia appear as projections from the hymenium (Figure 1.5). Basidia are four-pronged crowns shaped like miniature cow udders and bear a single basidiospore on each spike (or teat). One after the other,

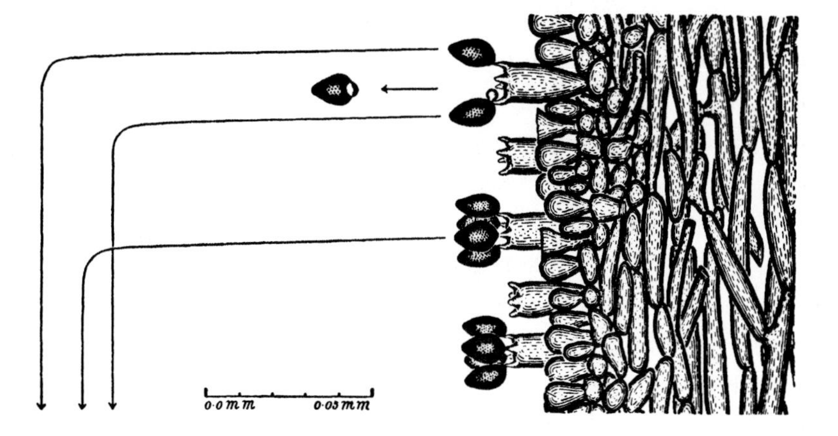

Fig. 1.5 Microscopic view of spore expulsion from surface of a mushroom gill. Note that the fluid drop is carried on the surface of the discharged spore. From A.H.R. Buller, *Researches on Fungi*, vol. 2 (London: Longmans, Green, 1922).

each of the spores in a quartet is catapulted horizontally, but only for a short distance so that it does not hit the neighboring gill. After this millisecond journey, gravity assumes control of the flight path, and the spore turns abruptly and accelerates straight down between the gills. Once the spore falls beneath the cap it is swept away by air currents. If an active mushroom is observed in the correct lighting, a dusty plume of basidiospores is visible swirling away from the cap (Figure 1.6).

Fig. 1.6 Cloud of spores dispersing from the horse mushroom, *Agaricus arvensis*. Mushroom illustrated in section to expose the gills. An immature fruiting body is connected to the same mycelium. From A.H.R. Buller, *Researches on Fungi*, vol. 1 (London: Longmans, Green, 1909).

The mechanism that catapults spores from the hymenium was solved only recently. It relies upon the condensation of water on the surface of the spore. A few seconds before discharge, a little bead of liquid develops at the base of the spore, grows until it becomes almost as wide as the spore itself, and then, instantly, fluid and spore disappear (Figure 1.7). John Webster, the egg hunter introduced earlier, tried to capture the discharge process using high-speed cameras at a film institute in Germany. The capacity for film wastage in this project was appalling. Webster watched through the microscope, holding a trigger for the camera and waiting for the appearance of the droplet of fluid. When the trigger was squeezed, thousands of frames of film were pulled through the camera in a couple of seconds by a deafening motor connected to the spool. But even at very high speeds, the best sequences showed hundreds of frames with a spore and its droplet, followed by hundreds of frames showing a naked spike of a basidium from which the spore disappeared.

Now that the catapult mechanism is understood, the photographic results are understandable. The spore is shot so fast from the gill that a camera running at 20,000 frames per second would be needed to capture the event.[7] The final speed of the spore is only one meter per second, compared with, for example, 7,800 meters per second for the Space Shuttle. But the acceleration of the spore is quite astonishing. From a standing start, this fungal cell covers a distance of one millimeter in a thousandth of a second. This sounds more impressive when you consider that the spore is only ten-millionths of a meter in length (10 μm), so that its journey corresponds to a distance 100 times its own size. Scaling up to human dimensions, this would be equivalent to vaulting from a cliff edge and almost instantaneously reaching a speed of 400 miles per hour. The spore pulls thousands of g's when it is flung from the gill, 10 times more than a jumping flea. This feat would atomize a bungee jumper.

The formation of the drop preceding spore discharge was first described by a French scientist, Victor Fayod, in 1889, but more than a century of research ensued before the discharge mechanism was explained by John Webster.[8] It is important to recognize that the space between the gills is saturated with water vapor that evaporates from the mushroom's tissues. Sugars and other molecules seep from the interior

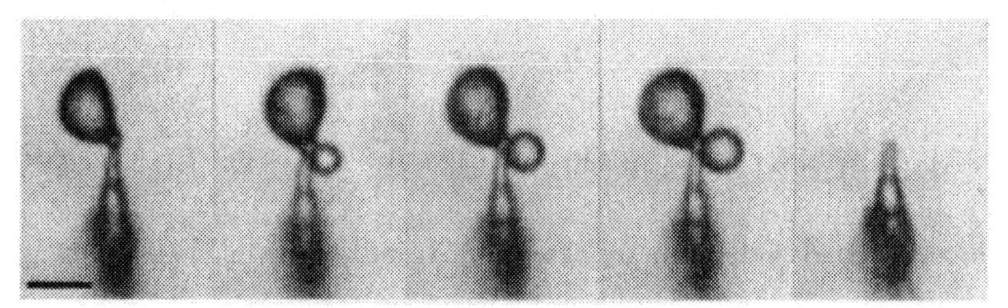

Fig. 1.7 Sequence of photographs taken at ten-second intervals showing the formation of the fluid drop at the base of the spore and subsequent discharge of the spore. The scale bar in the bottom left corner represents a length of 10 millionths of one meter (= 10 μm). From N. P. Money, *Mycologia* 90, 547–558 (1998).

of the spore, and their accumulation on its surface causes water to condense from this humid atmosphere. At the base of each spore is a hump called the hilar appendix, and water that collects here forms a spherical droplet that is held away from the rest of the spore. By this device, water condenses in two separate areas on the spore surface: one covering most of the spore, the other situated on its hump. Swelling of the drop on the hump pulls the spore's center of mass toward its base. (Similarly, if someone hands you a bucket of water, your center of mass moves toward the bucket.) Water continues to accumulate on the spore, and, finally, the two globs become large enough to make contact. When this happens, the drop rockets from the hump and merges with the fluid on the rest of the spore surface. You can get an idea of this attraction by watching two raindrops snap together when they make contact running down a windowpane. When this happens on the spore surface, the center of mass of the cell flies away from the gills in a millionth of a second, propelling the spore from its spike. (You would move too, if the tormentor that gave you the water, then kicked the bottom of the bucket, splashing you with its contents.)

A. H. Reginald Buller, professor of botany at the University of Manitoba, was responsible for much of the early research on the spore discharge mechanism, and bequeathed a treasure trove of original ideas to mycology when he died in 1944. Mycologists refer to the fluid drop on the spore's hilar appendix as Buller's drop in recognition of his work (Fayod did little more than note the drop's appearance). Some years ago, during an otherwise nightmarish sojourn in Kentucky (when feeling perilously close to becoming a murder victim was one of many nuisances), I found myself in a library fortunate enough to have Buller's seven-volume masterpiece entitled *Researches on Fungi.*[9] The books had not been checked out in more than twenty years, and I suggested to the librarian, that if shelf space was needed, I would be happy to buy them. The lady telephoned a few months later. For the cost of postage, nothing more, Buller's oeuvre was delivered to my door. One loose end discussed in Buller's *Researches* was the phenomenon of mushroom heating. A German mycologist, Richard Falck, had reported that some mushrooms could be up to 9°C warmer than the surrounding air, and thought that this warming might create convection currents that would assist spore dispersal. Without making his own temperature measurements, Buller contested Falck's conclusions, which stimulated me to revisit the heating phenomenon.

When I was hired as an assistant professor of botany, I was made aware that the acquisition of funding was more important than any combination of intelligence, creativity, and personality. Even the ability to teach is no longer an indispensable skill for the modern academic in some universities. To meet this professional mandate while preserving broad interests in mycology, I arrived at a workable strategy. This involved an A-list of research projects and a B-list. The A-list concerned research that had some probability of receiving government funding. The B-list drew on a Pandora's box of unconventional ideas that would provoke merriment among the officers of the National Science Foundation, followed by the dispatch of an icy rejection letter. Measuring the temperature of mushrooms—something that would get anyone other than a mycologist certified—qualified as a perfect B-list project.

Mushroom temperatures were taken by inserting thin wire thermocouples between gills and spines, or up inside the tubes of boletes and other fruiting bodies with pores beneath their caps. A group of students became deeply involved in the work, and measured temperatures from the hymenial surfaces of eighteen types of mushroom in the beech-maple woodland surrounding my university campus in Ohio. Contrary to the earlier work, we found that the mushrooms were colder than the air. I couldn't accept this result for the first few days and kept looking for errors in our method. But the thermocouples were very accurate, and we confirmed that mushrooms cooled during periods of spore release. Cooling was most extreme on warm days, when gills chilled as much as 5°C, but it continued even on cold mornings in November. We then studied mushrooms grown in the laboratory and found that cooling was stimulated by the passage of air around the fruiting body. This suggested an evaporative mechanism of cooling—the origin of the chill we experience after swimming.

The warming measured a century ago by Falck was probably due to the decomposition of mushrooms that he had plucked from the woods and brought into the laboratory. I'd never noticed how cold mushrooms felt before our experiments. Now I can't walk by a fruiting body without touching its cap. Mushrooms often feel very cold. You can test this for yourself the next time you are in the woods. Pluck a fresh-looking fruiting body and press its cap against your cheek. Some feel slimy, most are quite cold. *Xerula radicata*, the rooting shank, forms an elegant fruiting body at the base of deciduous trees. On a warm summer day, the cap of this fungus can feel positively frigid. Although we were the first to think of sticking thermocouples into mushrooms, fungal frigidity was recognized much earlier:

> they are all very cold and moist, and therefore do approch unto a venomous and murthering [murdering] facultie, and ingender clammy, pituitous [characterized by an excess of mucus], and cold nutriment if they be eaten.
>
> —John Gerard, *The Herball or Generall Historie of Plantes* (1636)

At first sight, cooling does not seem logical in relation to spore release because it appears at odds with the proposition that the condensation of water on the spore is the key to the catapult mechanism. But mushroom cooling makes more sense when we remember that the surface of the spore is sugary and differs from the basidia and all of the other cells of the gills. While water vapor streams from most of the mushroom's tissues, a little condenses on the spore, hydrating those sugars and making the droplet appear. The spore is comparable to a cube of bath-salts in a steamy bathroom. Steam fills the room when the shower is running, and water vapor condenses into liquid when it contacts the cube. This is why bath-salts crumble after a while even if they haven't been splashed with water. The spores and the bath-salts share this water-grabbing, or hygroscopic, behavior. Cooling promotes the buildup of water on the spore surface by slowing the movement of water molecules, encouraging them to occupy a more condensed state (liquid) rather than remaining as a gas (water vapor). The effect of temperature on condensation is familiar to anyone who has held a cold soda can on a warm day. If the mushroom warmed, even a little, water would evaporate from the spore surface, Buller's drops would never expand, and the spores would stay on the gills. Our experiments exposed the forest floor as a thermal mosaic, with mushrooms as its coldest inhabitants.

After mushroom spores are shot from the hymenium, they fall through the air spaces between the gills and then emerge from the lower surface of the cap and disperse in the air. It is surprising that so few of them become trapped inside the fruiting body because gills can be separated by less than 0.2 millimeters, and the tubes of some boletes and brackets are only 0.1 millimeters in diameter. Spore loss due to impaction on these surfaces is minimized by a number of adaptations. The catapult mechanism does not impart enough momentum to the spores to propel them onto an adjacent gill or spine, or onto the opposite side of a tube. Second, the cap and gills are highly responsive to gravity and orient themselves to provide unimpeded pathways for the free fall of discharged spores. But the shape of the mushroom may also assist in dispersal. This idea is supported by another B-list project in which we measured airflow patterns around mushrooms and models of mushrooms in a wind tunnel. Buller would have loved the idea of putting mushrooms in a wind tunnel.

The mushroom stem supports the cap above the boundary layer of still air on the ground (a few centimeters in depth at low wind speeds), so that the released spores will be swept away by wind (see Figure 1.6). But if the spores were exposed to wind immediately after falling from the cap, it would seem certain that many would be blown straight back into the cap, where they would become stuck on the bottom of the gills. In the wind tunnel it was apparent that when mushrooms are exposed to wind, the airstream divides at the leading edge of the cap and accelerates as it flows above and below. This airflow pattern is reminiscent of the aerodynamic behavior of an aircraft wing. Indeed, the air pressure beneath the cap exceeds the pressure above the curved surface of the dome, producing lift, and at high wind speeds the fruiting body can be wrenched from the soil. For a mushroom, this is an unavoidable hazard of life above the boundary layer. But this interruption of airflow also slows air movement immediately beneath the fertile tissues, so that the spores fall through calm air for a few tenths of a second. This may reduce the number of spores that are blown back into the cap, promoting dispersal.

Mushrooms with campanulate or bell-shaped caps are particularly effective at slowing the flow of air. There are many species with this shape that are most common in meadows, lawns, and other open locations. An impressive example is *Coprinus comatus*, a type of ink-cap called the lawyer's wig or shaggy mane. Young wigs are white and stand taller than stinkhorns. They recall photographs of ballistic missiles streaking from their silos. As this mushroom ages, it begins to blacken at the bottom edges of its gills and the elegant bell begins to resemble a ragged flag hanging from its pole. This species was the probable inspiration for Shelley's description of mushrooms in his poem "The Sensitive Plant" (1820):

> Their moss rotted off them, flake by flake,
> Till the thick stalk stuck like a murderer's stake,
> Where rags of loose flesh yet tremble on high,
> Infecting the winds that wander by.

Species of *Conocybe* grow on lawns and have much smaller fruiting bodies with bell-shaped caps. They are listed as poisonous and hallucinogenic, a profoundly dangerous combination. The prize among counterculture afficionados is *Psilocybe semilanceata*, the liberty cap, a mushroom with the same overall shape but decorated with a nipple at the apex of the dome. I prefer the mind-altering substances in a cup of Starbucks® or bottle of something blood-red and bone-dry, but I do remember a time when I was amazed that any professional mycologist would not indulge in the hallucinogens offered by his or her research subjects. But I digress. In addition to the manner in which these caps interact with air currents, the ubiquitous umbrella form of the mushroom is essential because it maintains a humid atmosphere around the hymenium while shielding the spores from rain. Any fluid running over the gills would wash away the fluid drops that develop on the spores and spoil the discharge mechanism. There are many ways, then, in which the shape and the physiological behavior of the whole mushroom contribute to its success at creating spore clouds.

The spores of gasteromycete fungi are spherical or ellipsoidal with no hilar appendix to mar their radial symmetry. The lack of the hump accords with the absence of the catapult mechanism, but evidence that gasteromycetes evolved from mushrooms that launched spores from their basidia is very strong. The fruiting bodies of a few gasteromycete species mature beneath the soil surface. These are called false truffles, and are attractive lures for rodents, which disperse their spores for the nutritional reward of the gasteromycete's flesh. Genetic data show that some false truffles are closely related to boletes.[10] The fertile tubes of boletes develop in the embryonic stage of the fruiting body beneath the soil and become exposed as the mushroom surfaces and expands its cap. Very few genes specify the emergence of the fruiting body and the bolete remains buried when these mutate. It seems that an important theme in the evolution of false truffles and other gasteromycetes is the gradual loss of the unfolding capacity of the fruiting body. This results in the production of spores within enclosed hymenial tissues and precludes discharge by the catapult mechanism. Fruiting bodies of gasteromycetes, like those of other mushrooms, are most plentiful after rainfall, but after they have expanded, many of these fungi can function perfectly during drought. As I have explained, spore dispersal depends upon animal vectors in phallic mushrooms, cages, and false truffles, and puffballs and earth-stars expel their spores in response to any kind of disturbance. The loss of the water-dependent mechanism of spore propulsion enables gasteromycete fungi to colonize much drier habitats than other basidiomycetes, and many of them are found on sandy soils and even in deserts. For these fungi, water conservation makes more sense than evaporative cooling, and not surprisingly, the temperature of phallic mushrooms, puffballs, and earth-stars is the same as the air that surrounds them.

Popular culture presents fungi as rather degenerate organisms, nasty growths that thrive on crap and corpses. A second representation introduces fungi, invariably a cluster of mushrooms with spotted caps, as a backdrop for wood elves, fairies, and pixies. In this chapter, by way of an introduction, I've shown that such illustrations do not begin to tap the strangeness and sophistication of the fungi. This book is also concerned with the scientific significance of these organisms, and it is difficult to underestimate their impact on the biological history of the planet. Human history is eclipsed by the succession of fungi that preceded our appearance. Fossilized hyphae of basidiomycetes have been dated to the Permian period (290 million years ago), and exquisitely preserved mushrooms have been found in Cretaceous amber from New Jersey. Although the tiny fruiting bodies of these fungi were immortalized in tree sap more than 90 million years ago, they are so similar in appearance to certain living mushrooms that they are instantly recognizable to a mycologist. They look almost identical to living species. Careful examination of the fossils reveals their spores stranded in the amber beneath the gills. When the amber is fractured, spores along the fault line are extracted to one or the other half of the jewel, leaving a perfect impression on the other side. The electron microscope reveals a projection at the base of each spore's footprint—the decisive signature of the catapult mechanism. So while herds of dinosaurs trotted past a wounded tree, and pterosaurs wheeled in the sky, the "dew of heaven" was beading on a mushroom. And then a glob of tree resin suffocated the fungus and preserved an instant in history. Next to the garden shed where I have begun writing about my love affair with fungi, a cluster of white-capped mushrooms has sprouted at the base of a dawn redwood tree. The dawn redwood, *Metasequoia glyptostroboides*, is another Cretaceous species, considered a living fossil by botanists. Its trunk is weeping resin.

CHAPTER 2

Insidious Killers

To say, for example, that a man is made up of certain chemical elements is a satisfactory description only for those who intend to use him as a fertilizer.

—H. J. Muller, *Science and Criticism* (1943)

Humans are vicious organisms, at least the carnivores among us, dependent upon the slaughter and dissection of vast streams of terrified herbivores. While most fungi are vegetarians, at least 300 species dine on human tissues, and many more exterminate other animals. I count myself among the few humans who love fungi, truly, madly, deeply. But my passion does nothing to offset the repulsion I feel at the sight of fungal growth on food, on the walls of a damp house, and much worse, on a human being. If fungi can rot bone in a patient's leg, feast on someone's brain, or devour a child's face, is the naked ape king of the jungle, or the king's dinner?

Jack Fisher was a researcher in John Webster's lab when I was a graduate student. Jack isolated fungi from all kinds of places, christened them with identification numbers, and sold the cultures to a company interested in mining these biological riches for novel pharmaceuticals. Jack suffered from terrible migraines and worked at odd times, often leaving the lab after toiling for many hours when I arrived at 9:00 or 10:00 A.M. (frequently clutching a hangover). Jack came across some fairly nasty fungi, isolates from clinical specimens ranging from dermatophytes that caused ringworm to vaginal yeasts to lung-infecting pathogens. We used an instrument called a laminar airflow hood to limit cross-contamination of cultures and reduce our own exposure to spores, but with stacks of agar plates filling the incubators and tottering on every patch of bench space, this was not a healthy environment. During the three years I spent at Exeter, I acquired a case of jock itch that resisted a battery of antifungal creams, and the feet of a fellow student, Nigel Hywel-Jones, were eaten by athlete's foot. Ever the opportunist, Jack took a scraping from Nigel's toes in case he had imported something interesting, but the fungus was identified as an escapee from one of the stacked plates. Nonhuman lab personnel made deeper sacrifices for Jack's partnership with industry. Nigel's tarantula, the late Mr. Terence, drew his brittle legs toward his hairy body one morning, slumped in a corner of his plastic cage, and began shedding clouds of spores. A rat rescued from the vivisectors contracted fungal pneumonia and coughed itself to death. This rodent belonged to another student, Joe Kirby, and enjoyed a rich diet of half-finished sandwiches and peeled grapes (he couldn't abide the skins). His airway infection was awful. Have you ever heard a rat coughing? He would lie on his back and hack away a whole morning, his little rib cage heaving up and down. At least he escaped the stainless steel guillotine operated by an animal technician and died among friends. Joe was suicidal at the loss of his lotus-eating rodent. Jack should have been arrested.

There is an important lesson in mycology here. All of the Exeter infections occurred in patients with healthy immune systems. The spores were simply too numerous and too varied in source. Nothing could resist this incessant barrage forever. There are many fungi capable of infecting otherwise healthy people, and this is a good place to begin this chapter on fungal diseases (or mycoses) of humans. Most benign are the infections of skin, hair, and nails caused by fungi called dermatophytes. Physicians refer to these diseases as ringworms, with Latin names indicating the infection site: tinea pedis for infections of the feet, tinea capitis for hair, and tinea unguium for nails. Hair and nails are composed of proteins called keratins that also comprise the outermost layers of the skin. Keratin is an example of an intermediate filament protein. It is extruded into hair follicles and then crystallizes, forming golden locks that cascade over her shoulders . . . excuse me, I was thinking about Michelle Pfeiffer. Now that you have pictured this screen goddess, consider that her hair represents a banquet for a dermatophyte. At least

1,000 calories are locked up in her keratin, equivalent to the energy in a couple of cheeseburgers. This nourishment is unavailable to animals that lack the enzymes needed to dissolve hair into an amino acid soup. We could do nothing but swallow hard and generate a very nasty hair ball.

In tinea capitis, the fungus invades the hair shaft, weakening the structure with an enzymatic vomit, and then reemerges on the outside of the hair and showers the scalp with spores. These spores can infect anyone's hair: yours, mine, or Michelle's. I'm scratching myself red as I write this paragraph. Dermatophytes are rare examples of fungi that are highly contagious.

Tinea corporis describes infections that spread over less hairy areas of the skin. Although distressing and often causing unbearable itching, these infections are rarely life-threatening. Ringworm fungi sometimes grow on a patient for decades and cause wholesale hair loss. They expand in ever-increasing circles, just as a mushroom-forming mycelium spreads in a meadow.[1] Tinea corporis can be manifested as an intricate pattern of concentric rings formed by overlapping scales of infected skin that eventually cover the entire body (Figure 2.1). From a distance, the patient looks tattooed. Fortunately, modern antifungal drugs are very effective at curing these infections. A dermatologist's portfolio of before and after pictures would paralyze a convention of witch doctors.

Even its most ardent critics must concede that science is not all bad; its application has allowed humans to achieve some remarkable feats, particularly in the United States where, ironically, a lunatic minority is bent on ridiculing both its method and its conclusions. The implicit message of religious fundamentalists intent on insulating children from current estimates of the age of our planet, Darwin's musings on biology, and other essential principles, is that the Western scientific tradition of truth by experiment does not work. To test the faith of these individuals I have designed an experiment. First, each zealot is sprayed with a culture of virulent ringworm spores and then, a few months later, the itching fundamentalists are offered a choice. Here is an antifungal medicine called Lamisil®, developed by some clever scientists at Novartis Pharmaceuticals, or perhaps you'd prefer to let this half naked and completely intoxicated gentleman wave a chicken over your ferocious

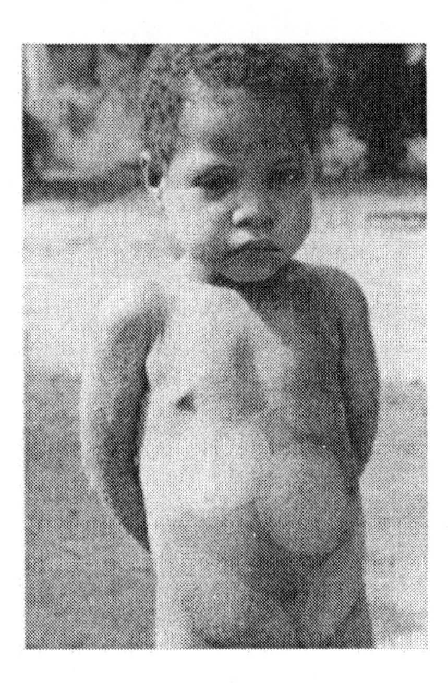

Fig. 2.1 Case of widespread ringworm photographed in Papua New Guinea. Photograph courtesy of Johannes Mattar.

lesions. Reason or groundless faith? Not all science is bad (just the big bits that tell us that we shouldn't look forward to an eternity of ice cream and great sex in a peaceful garden where we get to meet all our pets again).

Even the healthiest skin is colonized by a varied collection of fungi. The scalp is a grease bath where yeasts immerse themselves in sebaceous fat and gorge on dandruff flakes. Dandruff is a complex material. Scientists at the Procter and Gamble Company in Cincinnati are fascinated by it, a seemingly perverse interest explained by global sales of Head and Shoulders® shampoo posted at $1.2 billion per year. China is the biggest export market for this elixir (probably due to the dwindling supply of dodo semen or passenger pigeon eggs). Tom Dawson is a company biologist who introduces his research subject with an electron microscope image showing a single flake of dandruff. The morsel is a fragile raft held together by cells of the fungus *Malassezia*. *Malassezia* is a microscopic epicure. It can be grown on agar medium in a Petri dish, but only if it is bathed with olive oil. Tom swears by extra virgin. Dandruff is a complex complaint exacerbated by stress, diet, hormone balance, and even the weather, but it is always accompanied by *Malassezia*. The active component of Head and Shoulders® is pyrithione zinc, which inhibits fungal proliferation on the scalp, but the specifics of its mode of action are unknown.

Let's move from *Malassezia*, which everyone has, to *Madurella mycetomatis*, the fungus that causes mycetoma of the foot. Unlike athlete's foot, this infection can't be cured with a cream. Mycetoma is a tropical mycosis that begins with a splinter wound in a bare foot. Buried beneath the skin, the fungus grows for months or years, forming spherical abscesses and a series of interconnecting canals that resemble the tunnels in a termite mound. These canals or sinuses burst through the skin and weep a bloody fluid laced with infectious granules. As the disease progresses, *Madurella* erodes bones, producing a diagnostic "moth-eaten" appearance on an X ray. If the fungus does not become established in the head, neck, or chest (think about villagers carrying bundles of thorny firewood), patients do not die as a direct result of their infections. But by robbing an individual of unrestricted mobility, mycetoma can be a slow death sentence for someone in a developing country.

Madurella seems well adapted as a parasite. It forges a long-term relationship with its host, and can be dispersed from one foot to another via its infectious granules. Other fungi make better killers but poorer parasites. For these microorganisms, entry into human tissues can mark the beginning of the end for both host and pathogen. A yeast called *Cryptococcus neoformans* causes meningitis in about 10 percent of AIDS patients (Figure 2.2 a). This fungus is very widespread, and we probably encounter its cells on a frequent basis. When yeast cells or spores of *Cryptococcus* are inhaled, they are usually removed in the lifelong river of mucus that cleanses the deepest recesses of our lungs and then, by way of the throat, pours into the acid bath of the stomach. But when critical defenses are damaged in the immune-compromised patient, *Cryptococcus* travels from the lungs to the nervous system. In the brain it forms abscesses that lead to debilitating headaches, and eventually to blindness, dementia, and death (Figure 2.2 b).

Cryptococcosis is diagnosed by collecting samples of cerebrospinal fluid from a lumbar puncture. The clear fluid in the spinal column cir-

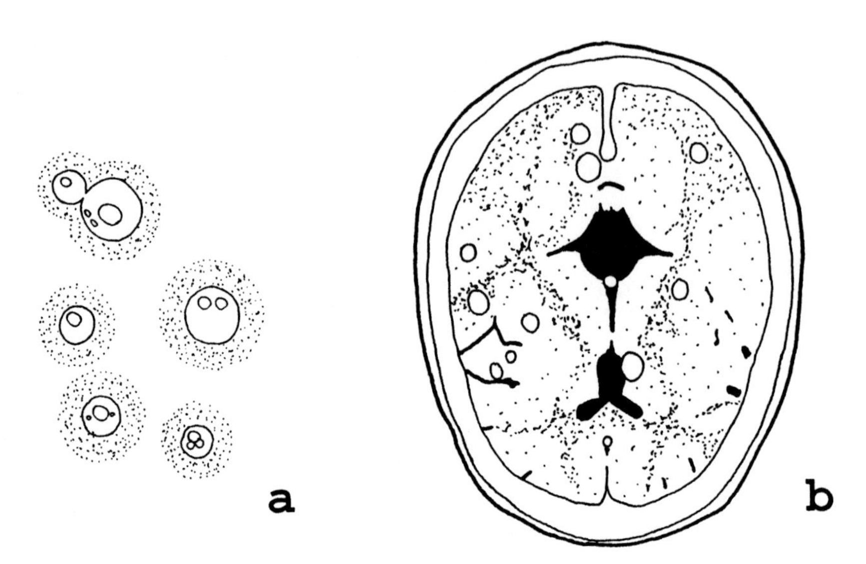

Fig. 2.2 (a) Yeast cells of *Cryptococcus neoformans*. Each cell is surrounded by a polysaccharide capsule. (b) Drawing traced from a CT scan of a brain colonized by *Cryptococcus*. The dark areas are the fluid-filled ventricles. The large circles scattered throughout the tissue are fibroid lesions called cryptococcomas which contain masses of the yeast cells. Drawing modified from image in A. Casadevall, & J. R. Perfect, *Cryptococcus neoformans* (Washington, D.C.: ASM Press, 1998).

culates all the way through the ventricles of the brain, and if a few drops of the sample from an infected patient are spread on a culture plate, colonies of *Cryptococcus* will appear in a day or two. Swifter diagnosis can be performed by direct microscopic examination of the cerebrospinal fluid or by using immunological tests that detect polysaccharides (large molecules built from sugars) produced by the fungus. I once dated a massage therapist who specialized in manipulating the movement of spinal fluid. Supposedly, there is a slow pulse associated with this internal canal and its activity reflects one's mental and physical well-being. Sitting in my garden shed on this frigid winter morning, I would welcome the touch of her warm hands.

Cryptococcal infections can be treated with high doses of a drug called amphotericin B, delivered as an intravenous infusion, or in extreme cases, injected directly into the brain (a treatment referred to as intrathecal administration). Amphotericin is a wonder drug synthesized by a bacterium that was discovered in a sample of soil collected from Venezuela in 1956. It acts against fungal cells by binding to a lipid molecule called ergosterol that sits in the plasma membrane. Like the cholesterol in animal cells, ergosterol controls the fluidity of membranes. Interference with this activity is lethal: amphotericin causes the plasma membrane of the fungus to perforate, and the debilitated cell begins to leak salts and sugars that are vital to its survival. Ergosterol is one of the few unique drug targets that are present in fungi but absent in humans, which sounds like a bulletproof sales pitch for amphotericin B. Unfortunately, when used at high doses for long periods of time, the drug causes side-effects that range from chills and vomiting, to anemia and kidney damage. But these disadvantages must be viewed in relation to the palliative nature of this chemotherapy for patients with advanced brain infections; when there is no expectation of a cure, the aim shifts to the relief of symptoms.

Amphotericin is often prescribed in conjunction with a second antifungal drug, 5-fluorocytosine. This compound impairs RNA function and DNA synthesis in the fungus, and also produces its own suite of side-effects. But fortunately, the action of the two drugs together is synergistic (greater than the additive effects of each when they are given alone). This combination therapy provides the same cessation of fungal growth with lower doses of each drug, diminishing the undesirable consequences of the treatment. Fluconazole is a third drug that is effective at halting the progression of a cryptococcal infection. It is an example of an azole antifungal agent, a drug class that blocks ergosterol synthesis by inhibiting an enzyme called 14-α-demethylase. Before the 1950s, death was swift and certain for a patient with cryptococcal meningitis. With prudent use of modern drugs many cases of the disease in otherwise healthy patients can be cured, and by suppressing growth of the fungus, the suffering of AIDS patients can usually be alleviated.

In July 2000, the International AIDS Conference in Durban, South Africa, provided scientists with a harrowing picture of HIV infection in the developing world, including the prevalence of cryptococcal meningitis. The tragedy of terminal AIDS in a hospital bed in the United States, with clean sheets and the distraction of a television, cannot compare with the nightmare of the poverty-stricken, HIV-positive African patient with an overpowering headache. Fluconazole has become one of the drugs of choice for treating cryptococcal meningitis because it is effective as an oral medication. At the time of the conference, a single daily dose of fluconazole, patented by the pharmaceutical company Pfizer under the brand name Diflucan®, cost about $10. This prescription charge is well beyond the means of most people in South Africa and other countries devastated by AIDS, condemning someone with fungal meningitis to blindness and death within a couple of weeks. A year later, to the amazement of some of the harshest critics of the pharmaceutical industry, Pfizer made the groundbreaking decision to offer an unlimited free supply of fluconazole to fifty of the world's poorest nations.

Medical mycologists do not regard *Cryptococcus* as a well-adapted pathogen. After all, it is rebuffed by most people unless their immune systems are damaged. For this reason, we wear latex gloves rather than biohazard suits in my laboratory when handling cultures of the fungus. It is a perfect example of an opportunist, a microorganism that ordinarily lives on dead tissues or feces—labeling it as a saprobe—but is sufficiently robust to tolerate occasional incarceration in a living host. At professional meetings, medical mycologists come close to expressing pity for opportunistic fungi. More time is spent discussing the hostility of the host than the virulence of the pathogen. Such benevolence is encouraged by the belief that once inside the human body the opportunist can never leave (not that this is any comfort for the host!). It's true that after *Cryptococcus* has entered the brain it is almost certain to drown in preservatives injected by a mortuary assistant, or roast in the flames of a crematorium. But these niceties of human disposal are very recent inventions. Bodies deteriorate rapidly in nature and cryptococcal cells might find their way into new hosts who contact the putrefying flesh or come close enough to inhale spores shed from its surface. The broad host range of the fungus argues for a lengthy evolutionary history: besides humans, cryptococcal infections have been reported in bats, camels, cats, cheetahs, civets, cows, dogs, dolphins, ferrets, foxes, goats, guinea pigs, horses, koalas, mangabeys, mice, pigs, rats, shrews, snakes, and a variety of birds.[2] Perhaps the yeast made its living in the brains of other hominids, millions of years before humans entered stage right, starring as Eden's bipedal blight. *Cryptococcus* cells possess too many adaptations that suit them for life inside animal tissues for me to relegate them to the status of reluctant intruders.

Speaking of decomposing flesh, I wonder whether there is any correspondence between the fungal species that evolved to rot animal carcasses and those that now infect patients whose defenses are compromised? It would be interesting to study the natural process of human decay from a mycological perspective, and I suggest that one of my professional colleagues with a strong stomach take a trip to the FBI facility in Tennessee called the "Body Farm." The resulting list of fungal decomposers would be intriguing. Supposedly, the mycelium of the basidiomycete *Hebeloma syriense* colonizes buried bodies, so that the appearance of its fruiting bodies can be evidence of a crime scene. Its common name is the "corpse finder."

June Kwon-Chung, a scientist at the National Institutes of Health, made an important discovery about *Cryptococcus* in 1976. She isolated fifteen cultures of the fungus from the tissues of infected patients and crossed pairs of the strains on agar plates by streaking their yeast cells together. Cells of four of the pairs became attached and fused. Fusion is cryptococcal sex, an interesting observation itself, but the next event was quite astonishing. From the glistening white colonies of merging cells, a different type of fungus emerged. Although it didn't resemble a mushroom, it grew as cylindrical hyphae rather than oval yeasts and formed basidia and bacterium-sized basidiospores (Figure 2.3). It was recognized as a relative of certain wood-decaying fungi with gelatinous fruiting bodies ("jelly fungi"), and was named *Filobasidiella neoformans*. When the basidiospores germinated, they produced the much larger yeast cells recognized as *Cryptococcus*.

Kwon-Chung's discovery showed that the yeasts which proliferated in the brain represented one growth phase of a fungus whose genes also encoded the instructions to make a spore-producing hyphal basidiomycete. *Cryptococcus* and *Filobasidiella* are different manifestations of the same species and a single genome. This type of relationship, called an anamorph-telomorph connection, is widespread among fungi. The yeast that grows in human tissues is the asexual part of the life cycle, termed the anamorph. These cells multiply by mitosis, the same way that yeasts proliferate in a bread mix. The basidiospore-producing fungus is the sexual stage, or teleomorph, whose development requires the union of two compatible yeast cells. This merger between **a**- and α-strains is akin to a fertilization event and equips the teleomorph with two copies of each chromosome. Later, a single copy of each chromosome is dis-

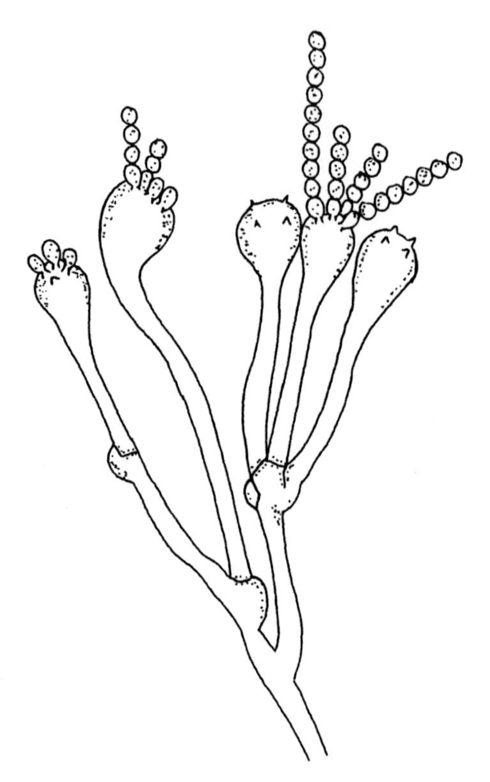

Fig. 2.3 Basidiospore-producing phase of *Cryptococcus neoformans* called *Filobasidiella neoformans*. Chains of spores produced by hyphae tipped by basidia. Drawing modified from illustration in K. J. Kwon-Chung, *Mycologia* 67, 1197–1200 (1975).

tributed to the nuclei of the basidiospores by a mechanism called meiosis. Animal eggs and sperm cells are generated by this kind of division. According to the formal rules for describing fungi, the name of the teleomorph has precedence over the name of the anamorph. So now that we understand the details of the life cycle, the name *Filobasidiella* should replace *Cryptococcus*, and be used when describing either phase. But it is not practical to scrap a name that is recognized by researchers and clinicians all over the world, and this strange situation, in which a single organism has two scientific names, persists.

Nobody is sure where the sexual stage of the pathogen exists in nature, although one group of investigators found its basidiospore-producing structures inside the flowers of eucalyptus trees. The basidiospores of *Filobasidiella* are smaller than the yeast cells of *Cryptococcus*, and are more likely to enter deep into the lungs if they are inhaled. This is one reason for pursuing studies on the dietary habits of the sexual stage of the pathogen. In an act of tremendous dedication to science, a German researcher prepared an aromatic culture medium called pigeon manure filtrate agar. Compatible strains of the yeast thrived on this concoction, mated, and formed the *Filobasidiella* stage. Experiments with pigeon manure were recommended by the fact that the fungus thrives in bird droppings, and researchers are intrigued by the possibility that pigeons may be a major source of the fungus in the urban environment. Bird droppings contain high concentrations of a nitrogenous compound called creatinine. Unlike many other fungi, *Cryptococcus* can utilize creatinine as a nitrogen source for protein manufacture, which may explain its predilection for growth in guano. While it would take a few minutes for a mycologist to convince someone that boiling pigeon droppings is a useful experiment, a photograph of a brain infected with yeast cells would probably convey the necessary message. I'm not sure that it would be as easy for the Hawaiian scientist who invented "dog-shit agar" in 1912 to justify his investigations.[3] I apologize if you have this book propped open on the breakfast table.

The sexual stage called *Filobasidiella* is formed by compatible strains of *Cryptococcus* that we refer to as the **a**-strain and the α-strain. These can be likened to males and females, although they look exactly the same under the microscope. Both strains can be isolated from the environment, but over 95 percent of all human infections are caused by the α-strain. This may be explained by the discovery that the α-strain is capable of making basidiospores without mating with the **a**-strain.[4] This maneuver provides the fungus with a shortcut for spore production: even if the α-strain is growing alone in pigeon droppings, it can bear infectious spores. Returning to my corpse-based model of infection, the body of a single person infected with this strain could also become a source of contagious spores, especially if the quarrying activities of insects exposed the infected tissues. This kind of broader historical and evolutionary perspective on human vulnerability may be helpful in understanding our relationships with opportunistic pathogens.

Medical mycologists are very interested in studying virulence factors, specific characteristics that enhance the ability of a fungus to cause disease. A conspicuous feature of *Cryptococcus* cells that enables them to live inside human tissues is the presence of a capsule built from mixtures of sugar molecules that extends as a wide halo around every cell (see Figure 2.2 a). This structure helps the fungus avoid detection and destruction by the immune system. Recently, there has been a great deal of research on the black pigmentation of cryptococcal cells. This coloration is due to the presence of a type of melanin, similar in chemical structure to the pigment in human skin. The pigment is deposited on the inner surface of the cell wall of the yeast, where it creates a resistant barrier. Although there is some uncertainty about the melanization process when the fungus is growing in human tissues, experiments on *Cryptococcus* cultures show that the pigment confers partial resistance to amphotericin B (demanding high doses of the drug), and can also protect the fungus from some of the defenses raised by the immune system. *Cryptococcus* can manufacture melanin from L-dopa, dopamine, or the hormone epinephrine, all of which are concentrated in the brain (L-dopa is used as a drug for controlling the tremors associated with Parkinson's disease). If L-dopa is added to nutrients in an agar plate, *Cryptococcus* can be distinguished from other microorganisms by its black colonies. Some mycologists believe that the blackening of the fungus explains why it colonizes the brain. It may attempt to grow in other locations in the human body, but in the absence of the precursors for melanin synthesis, its yeasts cannot become pigmented and are vulnerable to removal by the immune defenses. The idea that a fungus might steal chemical components from its host in order to protect itself from the host's defenses begins to indicate the insidious nature of human mycoses.

Other kinds of fungi that infect humans also synthesize melanin. The preceding sentence was as far as I penetrated for a number of hours, my mycological muse (Michelle Pfeiffer?) having fled my side, preferring to disappear into the woods across the street than stay with me in the shed. Other fungi that infect humans synthesize melanin—a different kind of melanin than the cryptococcal pigment, but still melanin. Then I looked away from the laptop and noticed a black stain on the cedarwood frame of the shed window. "Hello, who are you?" Just to be sure (and to do something other than writing for a while), I scraped some of the black stuff into a plastic baggie with my thumbnail. In my lab an hour later, I shook a little of the sample onto a glass slide, added a drop of water, and completed the sandwich with a coverglass. Magnified 400 times in the spotlight of a microscope, the dust resolved to millions of brown-black spores, clumps of them resembling grapes on the vine, each one fabricated from two or three separate compartments with a roughened surface. Within fifteen minutes, sitting in the sliver of warm water on the slide, the spores began to germinate. Translucent fingers, the organism's germ tubes, or hyphae, poked from fissures that opened in the black wall of the spores. Dried on a cedar windowframe just an hour earlier, the awakened microbe initiated a new mycelium. This is why I fell in love with mycology. It is a field of inquiry that can animate dust on a window frame.

Melanized fungi including my windowframe mold are ubiquitous and although they thrive as saprobes, some of them pose a serious threat to human health. They grow on surfaces wherever water is available, staining bathroom fittings, wallpaper, water bottles, coffee cups, and human beings. Soot from automobiles is often blamed for the discoloration of masonry, but the pollution is often fungal in origin. Melanized fungi coat the surface of buildings and ancient statuary, pit and penetrate the stone, and hasten its disintegration.

Fungal melanins are complex chemicals synthesized by the assembly of masses of ring-shaped molecules into an invincible polymeric web that will not dissolve in boiling water or hot acid. Deposited by the cell as a distinct layer of the wall, these pigments absorb all wavelengths of visible light (which is why the cells appear black), plus ultraviolet light, X rays, and gamma rays. The energy from these sources of radiation is transferred deep into the molecular structure of the pigment and is then emitted as heat, so that their damaging effects upon the living interior of the cell are diminished. Gentle warming of the cell is far preferable to corruption of the DNA. Resistance to ultraviolet light is one function of the pigment that allows melanized fungi to live on garden sheds and marble statues, while the translucent cells of other species are restricted to subterranean habitats. Part of the reason that lichens can survive in exposed locations at high elevations is that melanin within the hyphae of the fungal component of the organism protects the photosynthetic pigments of the algal partner. A greater testament to the resilience of melanized fungi comes from reports that dark-pigmented, radiation-tolerant fungi are flourishing in the contaminated soils around the Chernobyl reactor in Ukraine.

Much of our information on melanin's significance has come from experiments in which the behavior of a normal pigmented fungus is compared with that of an albino version. These translucent doppelgängers are created by poisoning the enzymes that the fungus uses to manufacture melanin, or by disrupting the genes that encode the same enzymes. In almost every contest between colored and colorless, the albinos are conquered. They are fine as long as they are grown in a stress-free environment on agar, but shine ultraviolet light on them and they shrivel, change the temperature in the incubator and they croak, or attack them with cell wall degrading enzymes and they explode. Melanin confers such diverse benefits that it is difficult to make a declarative statement about its function in any particular fungus, and competing laboratories have argued about the chief virtue of being black for decades. The only worthwhile generalization, which applies to brain-infecting fungi and Ukrainian molds, is that melanin furnishes the cell with a barrier between the cytoplasm and its hostile surroundings. It enables the fungal cell to rebuff poisons and prevent leaks.

To most of us, melanized fungi do not represent a significant danger, but even a black-stained shower curtain is a menace to someone with AIDS. Patients whose immune systems are compromised by cancer therapies or anti-rejection drugs prescribed following an organ transplant are also at risk. In rare cases, the same fungi colonize the tissues of patients with strong immune systems, but this requires trauma. Open heart surgery exposes a glistening smorgasbord for any fungal spore drifting through a hospital, and knife and bullet wounds, wood splinters, and accidents with nail guns also supply fungi with the necessary invitation for dinner. In common with *Cryptococcus*, a number of other melanized fungi institute themselves in the central nervous system. *Wangiella dermatitidis* is one of these rare causes of human disease that has a tendency to invade the brain. At autopsy, a brain slice from someone infected with this fungus with an impenetrable name (*huan-gee-ella, derm-at-eh-tide-diss*) shows tiny islands of yeasts intermingled with jet-black hyphae. Laboratory mice survive infections by albino mutants of the fungus, but perish when injected with the melanized strains. In culture, *Wangiella* switches from budding yeasts to tip-growing hyphae to penetrate agar, and may transform itself in the same fashion when it travels through solid tissues (Figure 2.4). In some human infections the nature of the trauma that admitted the fungus is a mystery, but there are other examples where the puzzle has been solved. A cluster of cases of *Wangiella* infections was tracked to a Japanese hot tub where a number of gentlemen had used a water nozzle to stimulate some exceptionally delicate tissues. Unfortunately for them, the melanized fungus also enjoyed the hot water and found itself driven into their skin by metal burrs on the surface of the nozzle.

Fungal infection following trauma is also encountered in diseases caused by unpigmented species. Zygomycete fungi are usually encountered as food-spoilage microorganisms. Far less fastidious than the fat-addicted dandruff fungus, zygomycete growth can be excited by an old bread slice or container of yogurt past its sell-by date. While some zygomycete spores are melanized, their hyphae are pellucid cylinders—beautiful cells that branch like candelabras and confine throbbing streams of organelles. When they're not growing on our food, one group of these fungi, the Mucorales, cause a family of infections called the mucormycoses. You would not wish one of these on your worst enemy.[5] Again, the disease can begin with something as simple as a splinter, but more often the person that develops a mucormycosis is already suffering from some underlying medical condition. Patients with uncontrollable diabetes are in a high-risk category, and infections are also seen in burn victims and chronic alcoholics.

We usually study hyphae in the simplified setting of an agar slab, and it could be argued that many of the things we have learned about fungi in the lab are next to useless when it comes to understanding how they grow in nature. But there is some overlap. The Mucorales grow astonishingly fast in agar, and they do the same thing in a human once they gain a foothold in its tissues. The nasal passages are a customary location for the mucormycotic mycelium, and by the time the infection is diagnosed it may be too late to save the patient. From the nasal sinus they are just centimeters from the brain, and they work their way in through tiny fissures in the bone or by following the walls of blood vessels

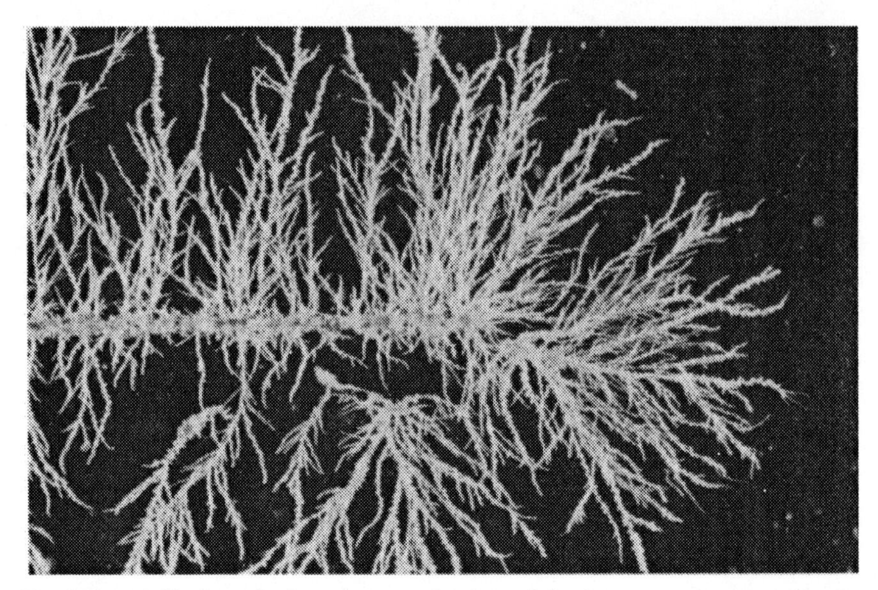

Fig. 2.4 Highly branched microscopic hyphae of the human pathogen *Wangiella dermatitidis* penetrating solid culture medium. A special type of illumination was used to capture this image so that the cells appear silvery against a dark background.

that pass through the skull. These microorganisms are classic opportunists, showing no obvious signs of the kinds of adaptation to human exploitation shown by *Cryptococcus*. They seem to use humans as a food source just as passively as they will eat a slice of pizza. Time for a panic attack: their spores are everywhere and, depending upon your location, you may have inhaled hundreds of them since beginning to read this chapter. Worship the miracle of your immune system. It is your greatest asset. Were he still alive, Howard Hughes (genius and world-class hypochondriac) would have paid me to stop writing at this point.

A possible case of mucromycosis of the lung was illustrated in 1855, and the first unequivocal description of cryptococcosis was written in 1894. Other human diseases caused by fungi have been recognized only recently, and we refer to some of them as emerging mycoses, those whose future disease-causing potential is troubling. I have spent the past few years studying the pathogen that causes pythiosis insidiosi, a rare, but potentially incurable mycosis. Our enemy is an oomycete (pronounced *o-o-my-seat*) fungus, or stramenopile, something not all closely related to the mushroom-forming basidiomycetes. Hyphae of this microorganism colonize skin and underlying tissues, produce corneal ulcers, damage the intestine, and invade blood vessels (Figure 2.5a). My introduction to pythiosis came as a student when I read a series of papers that discussed the cause of a curious disease of horses known as swamp cancer.[6] Swellings were seen in horses' lower limbs that were caused by the metamorphosis of smooth canon bones into monstrous coral-like outgrowths. Black-and-white photographs in these articles depicted miserable animals standing alone in their paddocks. In addition to bone malformations, other presentations of the disease in horses can include lumpy skin lesions called granulomas, growth of hyphae in the walls of blood and lymph vessels, and infections of the lung and intestine. When their skin becomes inflamed, the tormented animals scratch themselves against trees or fence posts until the infected tissues are exposed, leaving a wound that exudes a bloody serum as the underlying cells are consumed by the fungus.

Swamp cancer is one of many common names for the equine version of pythiosis insidiosi, which has also been called bursattee (also spelled burusauttee), Florida horse leeches, espundia, and summer sores. The earliest scientific descriptions of the disease were written by veterinarians serving with the British army in India. Addressing the editor of *The Veterinarian* in 1842, Charles Jackson, 8th Regiment of Light Cavalry, wrote: "I purpose, therefore, sending you from time to time a few lines on the diseases to which the horse is liable in India, which you may publish . . . or throw behind the fire, as your judgement or convenience may dictate." This attitude is amazing to those of us schooled in the contemporary publish-or-perish culture of academia. Jackson called the disease burusauttee, a term that refers to rain and the belief that pythiosis is associated with the monsoon season. He discussed treatment options, recommending oral cantharides, excision of infected tissue, and application of iodine solution to the wound. Cantharides are no longer used in veterinary medicine. They were prepared from the wing cases of the Spanish fly (actually a beetle), which are rich in chemicals called lactones. Today, Spanish fly is sold as an aphrodisiac. Other army veterinarians schooled in England were fascinated by the Indian mycosis. A 12th Royal Lancer, F. Smith, served in Madras and described the microscopic appearance of hyphae in scrapings from ulcers in *The Veterinary Journal* in 1884: "They converge and diverge, run parallel and transversely to each other,

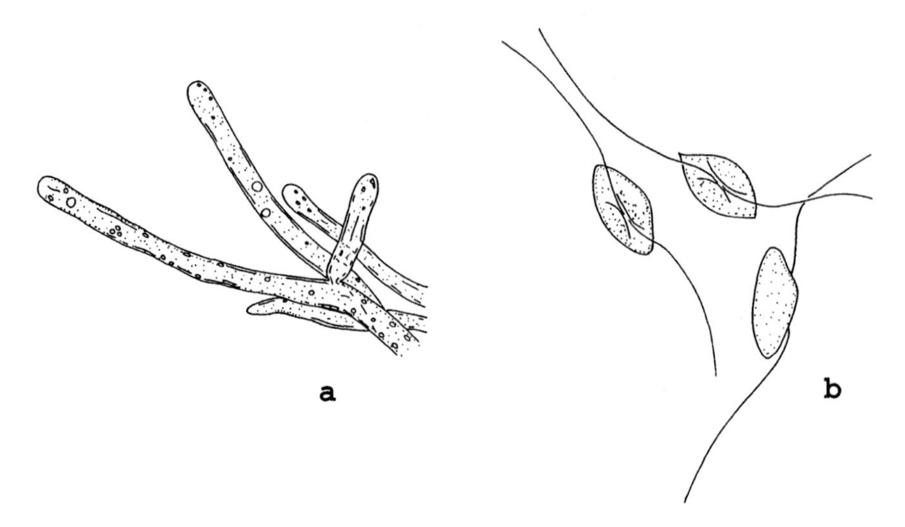

Fig. 2.5 (a) Invasive hyphae and (b) infectious zoospores of *Pythium insidiosum*.

communicate freely by branched processes, and some filaments appear curly." He had found the causal organism, but his findings were virtually ignored for a century. And then two cases of human pythiosis were reported in Thailand in 1987. Not only did the observations made by Jackson and later by Smith avoid the fireplaces of their editors, but these contributions to science survived to be cited more than a century later in articles that named their fungus *Pythium insidiosum*.

Pythium may grow as a soil saprobe, but it has not been isolated from anywhere other than mammalian tissues. In common with other oomycetes, *Pythium* produces sporangia from which swimming cells called zoospores are discharged into water (Figure 2.5 b). Under the microscope, young sporangia look no different from hyphae, but their character is revealed when they swell at the tip and discharge their contents into a spherical bag. Within a few minutes, the cytoplasm inside the bag is cleaved into packets of cytoplasm containing single nuclei. Each of these is fashioned into a single kidney-shaped zoospore with two flagella projecting from a cleft in its middle. Finally, the zoospores dart away through the water at great speed. In the laboratory, zoospore formation can be induced by transferring starved cultures to distilled water. But the natural process is a mystery because we don't know where *Pythium* releases its spores. The next stage in the development of the fungus is a little clearer: the zoospores infect a mammal.

A plausible story of disease transmission can be established. With the backdrop of a rainbow after a morning storm, a horse plods across its paddock toward a patch of taller grass, sprinkling droplets of water over its lower limbs with the imprint of each hoof. Each water droplet is airborne for a few hundredths of a second, carrying a fragment of the community of soil microorganisms that lay beneath the hoof. The droplet is a biological gem, containing soil bacteria, protozoa, nematode worms, plus one or two *Pythium* zoospores that rocket around and bounce off the inner surface of the drop. And then the droplet lands on the horse and flattens; water spreads between the horse's hairs, pressing the dab of microorganisms against the skin. Often the zoospores are splashed back to the soil or dehydrate on the horse's skin. No more hyphal growth or spore formation, the end of a promising career. But once in a while, an opportunity presents itself. Perhaps the horse nicked its skin on a thorn bush last night, or cut its leg on a spool of barbed wire. To the zoospore, this skin wound is the equivalent of a canyon full of chili. The carnivore stops swimming, jettisons its flagella, becomes sticky, and prepares for penetration. Within a few minutes, a slender hypha emerges from the spore and pushes into the underlying tissue. Pythiosis has begun. Soon, the mycelium is thoroughly embedded in a wet, protein-rich environment, kept warm by the obliging mammal. Think about bathing in egg yolks.

Contact with water containing the spores of this *Pythium* is a bad idea for any mammal. Besides horses and humans, pythiosis has been reported in dogs,[7] cats, cattle, and in one unfortunate polar bear in a zoo. There have been very few cases of human pythiosis in the United States; the disease seems to be more prevalent in Thailand and other Asian countries. One strain of *Pythium* studied in my lab was originally obtained from a male patient from Texas. A spider had chomped on his ankle, and the bite then became infected with the fungus. For the sake of a good story I'd like to say that this patient often walked barefoot and loved horses, but this is fiction. The zoospores may have been on the spider's mouthparts. In another case in Texas,[8] a boy developed an infection in his eye orbit after his sister elbowed him in the face (Figure 2.6). We must assume that the boy came in contact with water containing zoospores sometime after

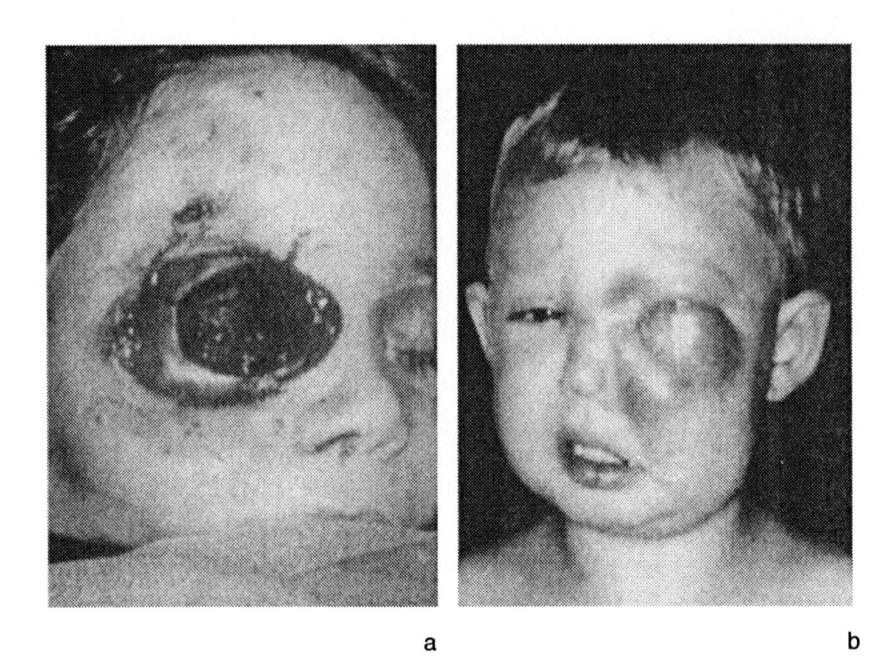

Fig. 2.6 Human pythiosis. (a) Infant with orbital pythiosis, treated by surgical debridement of infected tissues. Patient had sustained a mild facial injury preceding the infection. From E. S. Beneke, & A. L. Rogers, *Medical Mycology and Human Mycoses* (Belmont, Calif.: Star Publishing Company, 1996). (b) Inflammation and swelling of facial tissues of a 2-year-old boy infected with *Pythium insidiosum*. There was no evidence of prior injury in this case, but the boy had been playing in muddy water. From J. L. Shenep, et al., *Clinical Infectious Diseases* 27, 1388–1393 (1998, with permission). Both patients recovered from these infections.

the injury delivered by his sister. The boy survived following repeated surgical debridement of his infected facial tissues. Surgical debridement sounds nicer than carving away a child's face, but this is the reality of such a medieval therapy. Mealtime and misery for predator and prey. It is interesting to speculate that like *Cryptococcus*, *Pythium* is an ancient pathogen that has been here for millions of years, and probably for hundreds of millions of years. Although we serve as a contemporary food source for the pathogen, it is possible that the same microbe consumed the flesh of several kinds of prehistoric animals long before the appearance of mammals. Martha Powell, a world-renowned expert on zoospore-forming fungi, has found oomycetes growing in skin lesions on turtles and other reptiles in Alabama. A little imagination takes us to a lost world of exotic mycoses, a time when *Pythium* learned its flesh-penetrating tricks in the leg wounds of sauropods.

As I have mentioned, pythiosis can be an incurable disease. The fact that *Pythium* is not a close relative of mushroom-forming fungi is reflected in its chemical composition. Its membranes lack ergosterol, so the usual drugs that target this molecule are ineffective at halting the progression of pythiosis. In a few cases, these standard medicines have been prescribed and the patient has survived, but perhaps the immune system would have arrested the infection whether or not the drugs were delivered. One drug target deserves closer scrutiny. In his letter to *The Veterinarian* in 1842, Charles Jackson included iodine in a list of treatments, which is not surprising given its antiseptic properties. But interestingly, one of the first indications that oomycetes are not closely related to other fungi came by staining them with iodine. Microscopists found that the cell walls of oomycetes turned blue in an iodine solution, while the walls of other fungi did not bind iodine and remained unstained. This difference was due to the presence of filaments of cellulose (called microfibrils) that strengthen the cell wall of *Pythium* and its relatives. Perhaps iodine was a useful treatment for pythiosis in horses because it interfered with the synthesis of the *Pythium* cell wall. Currently, the use of vaccines is under investigation by Leonel Mendoza, a mycologist at Michigan State University. In addition to the successful treatment of pythiosis in horses, he cured a boy in Thailand with an arterial infection. The small number of cases of human pythiosis has limited acclaim for his work, but if cases of pythiosis become more frequent, Dr. Mendoza will become a superstar.

There is one important precaution necessary when working with *Pythium* in the laboratory: it must never be permitted to discharge its infectious spores. This isn't difficult if the surfaces of cultures on agar are kept relatively dry. Feeling confident in this practice, I have allowed a few keen students to work with the fungus. One afternoon one of my undergraduates brought a culture to my attention. Water had condensed on the inside of the plate and she wanted me to verify that the pathogen was growing under the misted lid. Holding the plate up to the light, water seeped from the edge of the plate and poured over my fingers. I told the student to pitch the plate and reminded her about the need for caution when handling cultures of pathogens. Without thinking too deeply about the possibility that zoospores had been swimming in that water, I washed my hands a few minutes later and resumed the familiar drudgery of grant writing. Weeks passed. Then looking in a bathroom mirror I noticed a cyst in my right eye. A fatty globule sat next to the iris. I rubbed my eye and went for a run. Half a mile from home, the digital image of the cyst intruded upon a movie concerning a wet culture plate that was playing on continuous loop somewhere else in my subconscious, and a pint of adrenaline poured into my bloodstream. I had done something extraordinary, something that would accord me a place in the annals of biology long after my death: I had infected myself with a lethal fungus. I pictured the removal of my right eye, debridement of the eye socket and deeper tissues as the surgeon tried to save the remainder of my swollen head. This bastard didn't synthesize ergosterol. I was going to die in agony!

My ophthalmologist, Geoffrey Collins, looked at my eye for a few minutes and said that I had a benign lymphatic cyst. For a few seconds I wanted to marry him. Then my heart rate returned to double digits and I went back to writing grants. We always wear latex gloves in my lab. We never rub our eyes.

I close this chapter by introducing a fungal disease that bears the name of my adopted home: Ohio Valley disease, also known as histoplasmosis. Given my revulsion for fungi growing on humans, I'm not too pleased with the thought that this organism may be sleeping in my lungs, but feel comforted with the strength in numbers argument which says that most Ohioans have been colonized by this species. *Histoplasma capsulatum* is another example of a fungus that loves bird droppings, and outbreaks of histoplasmosis have been associated with starling roosts. The sexual stage of this pathogen is an ascospore-producing villain that is named *Ajellomyces*. It was discovered by Kwon-Chung a few years before her breakthrough with *Cryptococcus/Filobasidiella*.

Ordinarily, our brushes with *Histoplasma* result in few if any symptoms, and if the fungus grows in the lungs it becomes encapsulated in little calcified nodules. But should the immune system lose power, the pathogen can get very nasty indeed, leaving the lung and penetrating any tissue in the body. Starlings are plentiful in my neighborhood this year.

A flock is chattering in the honeysuckle outside the shed window and splattering the snow with bright red berry stains. Holed up in here, surrounded by the enemy, I am Butch Cassidy to *Histoplasma*'s Bolivian army. When I bound from the shed and take that first breath of freezing air, the spores will enjoy the warmth of my lungs. For those of you smiling at the fortune of your geographic location, I have some bad news. Every day of your life other kinds of fungal spores pass into your nostrils and lungs. Given the opportunity, some of these may germinate and turn your body into soup. I have seen photographs of ink-cap mushrooms growing in a patient's throat, a little bracket-forming basidiomycete in a gentleman's nose, dead babies covered in yeast, vaginal thrush gone wild, and a moldy penis that infected my nightmares for a month. I would much prefer that fungi help decompose me after death, not before, although I recognize that few of us get to choose the method by which we make our ultimate contribution to the carbon cycle. I'm going to click SAVE now, close this file, and open the door.

CHAPTER 3

What Lies Beneath

Something deeply hidden had to be behind things.
—Albert Einstein, *Autobiographical Handwritten Note*

Like no other organisms, fungi flourish by burrowing into solid substances and transforming them into food. They penetrate the toughest leaves and woody plant tissues, skin, bone, and every other animal part. Some species even grow in granite. An ink-cap is eating a damp doorframe in my home, and making fungal growth an even more personal issue, I have a persistent itch between two of my toes (it could, of course, be much worse). All of us, itching or not, have good reason to be awestruck by the voracious appetites of fungi. Deliberately starved in the laboratory, a fungus that normally infects rice leaves will punch holes through bulletproof vest material in its futile search for nourishment.

Invasive growth by filamentous hyphae is a uniquely fungal process.[1] Before I explain how it works, it is necessary to describe the architecture of the mycelium and its component hyphae. I'll keep this as lively as possible, but recommend a shot of espresso if you feel sleepy at the first mention of ribosomes or cell walls.

Hyphae are eukaryotic cells, meaning that their chromosomes are housed in nuclei. Nuclei and many of the other subcellular structures or organelles found in hyphae are also components of human cells. The hyphal interior, or cytoplasm, is richly speckled with ribosomes. Ribosomes are the engines of gene expression that manufacture proteins. A high proportion of the billions of ribosomes in every cell stud an array of membrane-defined pouches—known as the endoplasmic reticulum—which is connected to the membranes that envelop the nucleus. Proteins forming on these ribosomes are extruded into the pouches, where they are snipped and decorated in preparation for their individual functions. Fungal mitochondria are sinuous organelles that function as furnaces in which food molecules are crumbled through the controlled chemical burn referred to as respiration. The energy from this process is captured in the form of smaller molecules that fuel all of the chemical reactions in the hypha. Most of the soupy cytoplasm is made of water, but it is suffused with a skeleton of protein strands that resembles a spider web in its multitude of interconnections and sensitivity to disturbance. Some of the strands of this internal cytoskeleton act as rails on which new cellular materials are brought to the growing tips of the hyphae.

The cytoplasm is contained by a plasma membrane that controls the entry and release of chemicals through integral proteins that act as channels or pumps. Some of the transported atoms and molecules carry an electrical charge (these are called ions), and their flux electrifies the membrane, turning the hypha into a microscopic battery. The importance of this energized condition will become clear shortly. By adjusting the exchange of chemicals between the cell and the external environment, the membrane regulates the mixture of dissolved salts, sugars, and other molecules that is essential for hyphal function. Finally, the outer surface of the plasma membrane is sheathed with sugar-based polysaccharides which are woven into a cell wall. This wall protects the hyphal contents from abrasion, screens out poisons, and, by restricting inflation of the cytoplasm, allows the cell to become pressurized.

Fungi grow by extending the tips of their hyphae and by branching to initiate new hyphae. As the organism expands, it forms a maze of filaments whose pattern bears some resemblance to a series of river deltas. This is the colony, or mycelium, the feeding phase of the fungus. Most often, hyphal growth is invasive, so mycelia develop inside the food source (or substrate) utilized by the fungus. In mushroom-forming basidiomycetes, the mycelium is submerged in the soil or rotting wood beneath the fruiting bodies; in pathogens, mycelia sink themselves in host tissues. An individual mycelium can occupy a very restricted volume (think of a ringworm fungus digesting a single hair), or become colossal.

If food is plentiful, the mycelium tends to adopt the shape of a doughnut or torus by spreading outward from its point of origin, leaving an ever-widening circle of dead and dying cells at its core. This distinctive growth pattern is responsible for the creation of fairy rings in pastures and lawns, which spawn annual flushes of mushrooms above the active perimeter of the mycelium. Large mycelia belonging to a basidiomycete called *Armillaria* have been discovered in the United States. The outer rim of the *Armillaria* mycelium, consisting of billions of hyphal tips, can travel through a forest for thousands of years, feeding on organic matter in the soil and invading tree roots. Because different fungi can mingle in the same general area, it is difficult to follow the development of individuals by examining soil samples. Fortunately, fruiting bodies duplicate the genetic makeup of their soil-bound parents, so by collecting and analyzing mushrooms, investigators can track the otherwise invisible expansion of the mycelium. In this way, it was discovered that the mycelium of a Michigan giant spread itself over 15 hectares (or 37 acres) and equaled the mass of a blue whale.[2] (Similar molecular genetic methods tracked the movements of O. J. Simpson in 1994, suggesting that he had not spent a quiet evening eating hamburgers with that winsome servant whose name we have all forgotten.) More recently, a monster *Armillaria* has been found in the Blue Mountains of eastern Oregon that may cover more than 2,200 acres, and I won't be surprised to learn someday that this too is eclipsed by something growing in the vast forests of Siberia. From the size of the mycelium and its current rate of expansion, the age of the Oregon fungus is estimated to be between 2,400 years and 7,200 years. Currently, *Armillaria* holds the title of World's Largest Organism, soundly beating its nearest competitor, the quaking aspen. Aspens also spread as colonies in the western United States, with hundreds of clones sprouting from a single tree. Some forest patches may be 10,000 years old, but they are outclassed by the fungus in terms of living mass, because dead wood accounts for much of the incredible bulk of an aspen grove.

Mycelia can be constructed in two different ways. The first method produces a colony of interconnected compartments (Figure 3.1 a). When cultures of these fungi are grown on agar and viewed with a microscope, the hyphae appear as railroad tracks with wide-spaced ties ("sleepers" in England). This is an optical illusion that renders the cylindrical wall of the hypha as a pair of parallel lines and recasts cross-walls

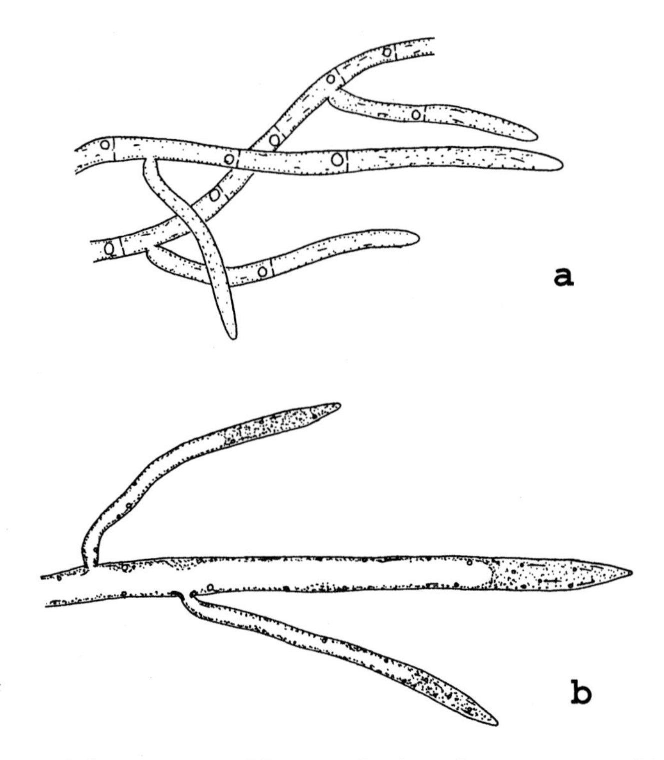

Fig. 3.1 Hyphal construction. (a) Septate hyphae of an ascomycete. (b) Branched non-septate hypha formed by an oomycete water mold. The central part of the cell is occupied by a fluid-filled vacuole.

or septa into ties. In reality, the septa are discs, and each is perforated with a simple pore, or an elaborate barrel-shaped valve capped by membranal domes. Close to the tips of hyphae, cytoplasm shuttles from one compartment to the next. Farther behind the extending tips, older compartments show less motion and are eventually sealed from the rest of the colony when their septal pores are plugged. The toroid form of the whole mycelium is a reflection of this programmed senescence, which occurs within every one of its constituent hyphae.

Because physical continuity exists between the fluid contents of adjacent compartments, we say that hyphae are coenocytes, single cells that contain many nuclei. Studying wood decay fungi in the 1980s, the late Norman Todd, a British geneticist, observed nuclei worming their way between compartments through septal openings. Fifty years earlier, A.H.R. Buller had postulated that nuclei must move rapidly through mycelia, but until Todd's work this nuclear traffic had not been seen. Norman's ghost just reminded me that I had said that there were two methods of hyphal construction. The basidiomycetes, and a second group of fungi called ascomycetes (featured in the following chapter), form septa. Hyphae of other fungi are not partitioned by these cross-walls. Oomycete water molds like the human pathogen *Pythium insidiosum*, and the zygomycete fungi have this non-compartmentalized organization (Figure 3.1 b). I mentioned zygomycete fungi in the previous chapter—some of them cause human infections called mucormycoses—and we'll encounter them again later in the book. Zygomycetes are distant relatives of the mushroom-forming basidiomycetes and the ascomycetes: all three groups are assembled within the same taxonomic group called Kingdom Fungi.

I often refer to members of this triad as "mushroom relatives" to distinguish them from the oomycete fungi that occupy a distinct kingdom of eukaryotes called the Stramenopila, which also incorporates diatoms, kelps, and other kinds of photosynthetic algae. Kingdoms tend to be very large groupings of organisms. All animals, from brine shrimp to humans are accommodated in a single kingdom, the Animalia, and all plants are members of Kingdom Plantae. Not surprisingly, then, the distinctions between the Fungi and Stramenopila are profound. But oomycetes have been treated as fungi since the origins of mycological study because they have invasive hyphae, feed by absorbing nutrients, and form spores when they reproduce. They are featured throughout this book, and although I'll refer to them as fungi, remember that they're not close relatives of species that generate mushrooms.

I have studied hyphae for many years and have chosen some favorites. The most beautiful are produced by a family of oomycetes called the Saprolegniaceae. These include *Achlya* ("the flower without petals"), *Thraustotheca*, and *Dictyuchus*, whose hyphae expand to a few tenths of a millimeter in diameter, considerably bigger than most other hyphae. When illuminated on a microscope stage, these cells sparkle, as parallel streams of organelles glide through their translucent cytoplasm. The pointed tips of the hyphae are crammed with cytoplasm, but behind the growing apices the cytoplasm forms a thin sleeve around a cylindrical structure called the vacuole. The vacuole is filled with water and dissolved ions and molecules and appears perfectly transparent. This pool is separated from the cytoplasm by a membrane, which is similar in composition to the plasma membrane that rests against the wall. The interior of the hyphae can be explored with a microscope; by changing focus, you can dive into the vacuole, "swim" up and down into branches, and (at least in my case) waste a great deal of time before embarking on something you promised a funding agency that you would pursue with the zeal of an evangelist. The Saprolegniaceae should be combined with the stinkhorns into the category of "mycological spectacles one should not miss." William Arderon, a Fellow of the Royal Society of London, described their hyphae in 1748 when he found a forest of filaments radiating from an infected fish:

> These fibrils, when examined by the microscope, shew themselves to be a number of minute tubes, filled with a brownish liquor; and this liquor, upon pressing them, becomes immediately discharged.
>
> —*Philosophical Transactions of the Royal Society of London* 45: 322

His published illustration shows the filaments of *Achlya*, or one of its family members. Arderon's description is a landmark in mycology because it was the earliest definitive report of a fungal disease in a vertebrate.

Life in the form of networks of branched hyphae has some interesting consequences. The plasma membrane is a continuous film, comprised of lipids and proteins, which lines the cell wall throughout the mycelium. Wherever septa are present, the membrane narrows and passes through the central pore, and then enlarges to line the wall of the next hyphal compartment. If the wall of a hypha were removed, the naked membrane would resemble a chain of sausages, pinched at every septum. In the absence of septa, the membranes of fungi like *Achlya* are shaped as uninterrupted cylinders. But whether they are septate or nonseptate, all mycelia are constructed as a series of interconnected tubes. Like an oil pipeline, this structure is vulnerable to catastrophic injury. When a hypha is damaged so that its membrane is ruptured, perhaps by a browsing worm or insect, pressurized cytoplasm surges along the entire length of the cell toward the wound. If the hypha is mutilated, the pipeline analogy works well: its guts spill freely. But fungi can repair less extensive damage with remarkable proficiency. Within seconds, hyphae stop small holes with fresh membrane, and crippled compartments (in septate hyphae) are isolated by sealing septal pores to protect the rest of the organism. Growth can then resume within minutes, sometimes by the formation of branches from either side of the wound.

Earlier, I mentioned that hyphae are electrically active. This is true of all kinds of cells, not just those of animal nervous systems or the bizarre tissues of electric eels. It is possible to measure the voltage maintained by a hypha by piercing its cell wall and plasma membrane with an instrument called a microelectrode (Figure 3.2). The microelectrode is a glass needle, or micropipet, with a sharp tip with a microscopic opening. It is filled with a salt solution and connected to an amplifier. The electrical circuit is completed by connecting another electrode to the amplifier and positioning its tip in fluid bathing the fungus. Movement of the micropipet is controlled with an instrument called a micromanipulator. As the pipet is eased into the hypha, the plasma membrane forms a tight seal around its tip. In a successful recording, the amplifier registers a hyphal voltage (or potential difference) of 100 millivolts to 200 millivolts. This voltage is sensitive to all kinds of perturbations. I know this because I've listened to the screams of hyphae.

At a scientific meeting in the 1990s, I met two mycologists from Europe who had perfected a method for measuring the membrane voltages of hyphae growing in bundles through soil or wood fragments. In addition to the usual connection of the amplification system to a chart recorder, the researchers converted the amplifier output to an audio signal so that changes in voltage could be heard as alterations in pitch. In the name of science, they then proceeded to torment the hyphae. The cells were allowed to dry out at one end (fungi cannot grow without a continuous supply of water); they were heated, cooled, and even severed with scalpels. Once I understood the description of their methods, headphones were placed over my ears and the researchers started their tape recorder.

Click . . . click . . . click . . .
Click . . . click . . . click . . .
Unt now ve are burning zuh fungoose . . .
Click, click, CLICK, CLICK, BUZZZZ!

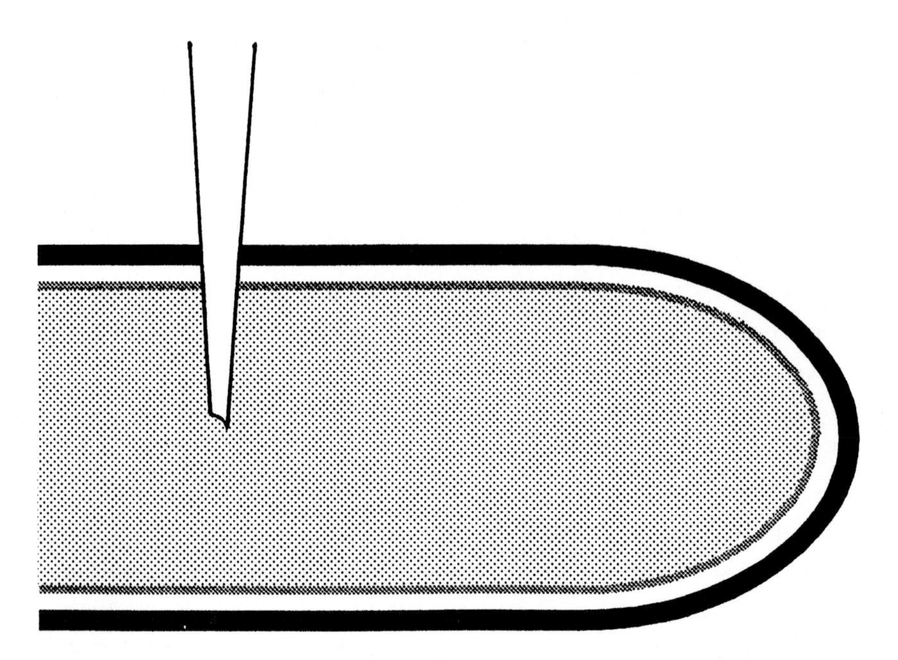

Fig. 3.2 Diagram showing hypha penetrated with glass micropipet. The micropipet passes through the cell wall and underlying plasma membrane of the hypha, so that its tip is situated in the cytoplasm. The voltage across the plasma membrane can be measured by connecting the interior of the pipet to an electrical amplifier.

CLICK, CLICK, click, click . . .
Click . . . click . . .
Now zu cells iz cut mit der scalpul . . .
Click, WHEEEEE!

Damage to one part of the mycelium seemed to affect cells far removed from the injury, demonstrating a surprising level of irritability for an organism that lacks a nervous system! Their macabre nature aside, these experiments illustrate something that mycologists have suspected for many years. Rather than growing as isolated strands, the connections among all of the hyphae in a mycelium allow these filamentous cells to cooperate as a single organism.

This concept can be understood by considering a mycelium that branches inside a decaying log, using cellulose (the cell walls of the dead tree) as its chief source of carbon. Such fungi are described as cellulolytic, because they break down cellulose into its component sugar molecules. As the cellulose disappears, the wood turns brown, because all that remains is a dark-colored polymer called lignin. This type of decay is classified as a brown rot. A rotting log contains veins of dense, nutrient-rich tissue, separated by crumbly areas that are already degraded, and intervening spaces filled with air or water. So in response to nutrient absorption by some hyphae and starvation of others, the colony becomes sculpted into a series of fans with an increasing number of hyphae growing toward nutrient-rich regions and few if any between them. The fungus can also move food and water between different parts of the mycelium to maximize the potential for increasing its mass. When enough food has been absorbed and when the environmental conditions permit, the mycelium will produce a flush of mushrooms. Sometimes a fungus can do this on its own, but usually sexually compatible mycelia must fuse with one another. We'll save fungal reproduction for a later chapter.

Hyphae feed by absorbing nutrients through their walls and plasma membranes. It's not clear whether they soak up nutrients over their whole surface or just absorb food at the hyphal tip. But it is likely that more food flows through the tips than elsewhere, because this is the main location where the fungus releases its polymer-degrading enzymes. Enzymes are proteins that accelerate the rate of reactions between chemicals. An astonishing array of enzymes is released by mycelia and there are very few materials that are impervious to fungal degradation. Cellulases are hydrolytic enzymes that employ water molecules to attack bonds between sugars in cellulose molecules, and proteases (or proteinases) act in the same way when they break proteins into their component amino acids. In damp environments fungi will grow on the coatings on camera lenses (and etch the glass), eat the plastic lining of fuel injectors (fungi are said to have grounded an Arab air force), pit the surface of machine tools, liquify wallpaper, and consume just about anything else given enough time. A black mold is working on a shampoo bottle in my shower, which is ironic because the contents of the bottle are supposed to possess antifungal properties that help suppress dandruff (this fungus is in for a surprise if it breaks through the plastic).

Fungi draw upon a seemingly boundless catalog of enzymes to digest their surroundings. These proteins are highly specific in their targets: a particular fat-degrading enzyme may operate against only one type of lipid molecule, at a particular temperature, at a specific level of acidity. My wife and scientific collaborator, Diana Davis, has been studying proteases for the past few years. She has found that a fungus which kills grass on golf courses and a related species that feeds on marine algae release massive quantities of these enzymes when they are fed animal proteins. Although they operate as vegetarians and never grow on animals in the wild, these fungi have an innate capability to consume animal tissues. This nutritional flexibility is astonishing. (Humans are described as omnivores, but even something as innocuous as goat cheese gives me indigestion.)

Their catholic tastes suggest two things about fungi. First, they are prepared to eat anything, and second they have had a very long evolutionary history. The second conclusion rests on the idea that parasitic or pathogenic fungi[3] specialized for growth on particular species of plants or animals probably evolved from saprobes that grew on organic debris like decaying leaves and rodent carcasses in the soil. The scavenging strategy of saprobes requires a level of biochemical dexterity not needed by pathogens that always consume the same things. But it appears that as some fungi evolved the finicky characteristics of disease-causing microbes, they retained the genes that encoded enzymes they no longer needed on a daily basis. The conservation of these genes suggests that they still use them once in a while, and it is true that most pathogenic fungi can grow as saprobes when they find themselves outside their hosts.[4] Through this strategy fungi have preserved an astonishingly broad inventory of enzymes.

The dietary breadth of fungi is an important issue for geneticists who have tried so hard and spent so many tax dollars in their quest for compounds that will halt the growth of pathogenic species. The philosophy behind this research seems logical. If we could identify an enzyme that is essential for fungal penetration of a particular host, then an antifungal compound that inhibits the action of this enzyme should obliterate the disease.[5] Plant leaves are covered with a waxy layer called the cuticle. Fungi attacking plants push through the cuticle before penetrating the second hurdle presented by the walls of the plant's epidermal cells. Fungi release enzymes called cutinases that dissolve the cuticle, and the suggestion was made that disruption of cutinase production might stall the fungus on the leaf surface and protect plants from infection. Years of research ensued, during which genes encoding cutinases were identified, sequenced, and finally disrupted to produce cutinase-deficient mutants. Sure enough, the mutants stopped producing the target enzymes, but when their spores were sprayed on leaves, they germinated, punctured the cuticle without any apparent difficulty, and ate the plant. The first problem with these experiments was recognized fairly quickly. Fungi produce many different cutinases, so that removal of one enzyme has little effect on the cuticle-degrading abilities of the fungus. The answer then, was to disrupt several different cutinase genes, and to knock out more than one of them in a single fungus to create ever more disabled mutants. But this strategy was also unsuccessful. Contrary to the original assumptions, the cuticle appears to be quite insignificant as a mechanical barrier to invasion by some fungi. By dissolving the cuticle into its component lipids, cutinases probably function as nutritional enzymes that allow the newly germinated fungus to obtain nutrients from the otherwise barren leaf surface. But the observation that the mutants push on and infect the plant without these enzymes shows that the young fungus does not starve without this hors d'oeuvre.

Thwarted by the cutinase experiments, researchers reasoned that by disrupting fungal genes that encoded enzymes which degraded the wall underneath the cuticle, they would concoct the long-sought mutants incapable of forcing their passage into the plant. The plant cell wall consists of cellulose microfibrils (which are also found in the walls of oomycete fungi) glued together with other polysaccharides and embedded in jelly. These materials comprise the bulk of a plant's tissues and surely offer more substantial deterrents to invasion than the waxy veneer on the leaf surface. All kinds of enzyme-encoding genes were disrupted, but the deviant fungi snubbed the scientists and continued killing plants. Again, fungal flexibility was an enemy of the researcher. To clear a path through the mixture of polymers in the plant cell wall the fungus probably releases a whole cocktail of enzymes. Strike two.

Hyphal invasion of human tissues is a second focus of funded research on fungi. Enzymes that dissolve plant cells are powerless against animal tissues, and proteases are believed to play the roles of barrier removal and food acquisition for human pathogens. Some of the most informative experiments have been performed on *Candida albicans*, a fungus described as a commensal, a normal component of the rich microbial community that lives on and inside our bodies. As a yeast it thrives on human skin and mucosal linings (e.g., the vagina, throat, and intestine), and up to 1,000 *Candida* cells can be present in every gram of feces. But often it becomes unruly, causing vaginal thrush, and its presence in the birth canal can lead to proliferation of the yeast on newborn babies. (Oral thrush, or aphtha, was once a major cause of infant mortality, and is referred to frequently by Dickens and other Victorian authors.) In rare cases the fungus becomes very aggressive, switching from a budding yeast to a hyphal form and digging into solid tissues. As we have seen with other mycoses, these infections are usually associated with damage to the immune system. Fluconazole is highly effective at treating different kinds of *Candida* infection, but recent encounters with drug-resistant strains have heightened the demand for new approaches to annihilating the fungus. Reasoning that a specific type of protease allows *Candida* to penetrate animal tissues, medical mycologists have disrupted the genes encoding these enzymes. But in a replay of the experiments on plant pathogens, they have found that the resulting mutants retain their expertise at killing laboratory mice.[6]

Most mycologists have now abandoned the conviction that we can elucidate complex mechanisms like tissue colonization by interfering with a few secreted enzymes: there is no magic bullet, no single tissue-degrading enzyme whose inhibition would stop the pathogenic fungus in its tracks. But there is great hope for future genetic experimentation. By taking account of the flurry of gene expression that accompanies the invasion of living tissues, a few laboratories have reported success in reducing the virulence of pathogenic fungi by disrupting regulatory genes that control the production of whole groups of secreted enzymes. This type of approach may provide investigators with the mastery over fungal development that has eluded them for so long. Technical innovations in the field of genomics and proteomics are even more exciting for mycologists. No longer limited to the analysis of a handful of genes and proteins, new methods will illuminate every protein manufactured when a pathogen plunges through a leaf, or when *Candida* is transformed from a yeast into a hypha. Comprehensive readings of the activity of every gene and gene product in the cells of the fungus and its host will revolutionize the study of the infection process.

Hyphal enzymes are produced in the cytoplasm and packaged in minute spheres called vesicles that bleb from the flattened pouches of endoplasmic reticulum mentioned earlier in the chapter. They stream toward the tip of the hypha and fuse with the existing plasma membrane, adding new surface to the elongating cell and releasing their contents into the wall. This is the process of secretion. Some of the vesicles' enzymes remain trapped in liquid-filled spaces in the wall, while others diffuse into the surrounding environment. For hyphae growing through wet logs, cellulose-degrading enzymes bind to cellulose and release sugars. Similarly, *Candida* bathes itself in amino acids when it grows in human tissues by secreting proteinases. Once sugars, amino acids, and other small molecules are released from the substrate, fungi use their electrical activity to absorb them.

In the 1980s, I spent two years studying a transport protein that exports charged hydrogen atoms or protons from the hypha across its plasma membrane. At the time we were engaged in a Frankenstein-like attempt to make frog eggs assemble fungal proteins. (Unfortunately, we didn't learn very much about frogs and discovered even less about fungi.) As the hypha spits protons, the flux of these ions charges the membrane and contributes to the voltage measured with a microelectrode. A healthy hypha lives far from the equilibrium condition in which the concentration of protons would equalize on either side of the membrane, and this unbalanced state drives the reentry of protons if an opportunity is presented. Hyphae exploit this thermodynamic imperative to import other elements and molecules along with the influx of protons. For example, there are carrier proteins situated in the plasma membrane that offer pathways for proton reentry and coincident uptake of sugars. The carrier functions by opening and closing in a specific fashion each time it transmits a hydrogen ion, so that it also shunts a sugar molecule into the cytoplasm. This is one of the processes that allows the cell to accumulate the food molecules needed to meet its energy demands. Like all living things, the fungal mycelium is an island of chemical order dwarfed by an expanse of disorder.

Chemical reactions that move materials through the plasma membrane of the hypha also establish a preferred level of internal saltiness, acidity, and hydration. Movement of water into the cytoplasm, termed osmosis, is one consequence of the accumulation of various chemicals by cells. Some organisms deal with the constant influx of water by pumping it out. Amoebae work in this way. But hyphae have a different strategy: they become pressurized. This is possible because the hyphal wall resists cytoplasmic expansion and as the membrane presses against its inner surface, hydrostatic pressure builds to a few atmospheres. Pressurized hyphae are described as turgid.

For over a century, mycologists agreed that hyphae use pressure to drive expansion of the wall at their pointed tips. The same idea is widely accepted as an explanation for the way that plant cells expand. As a researcher, I have promoted a somewhat unpopular view that turgor pressure is a red herring when it comes to making sense of growth. Some years ago, Frank Harold and I discovered that oomycete water molds actually grew faster when we eliminated most of their internal pressure. This was done by cultivating colonies in high concentrations of soluble carbohydrates and synthetic compounds to disturb their natural osmotic balance and limit influx of water. In two sentences I've summarized the results of seven years of technically challenging research, and I'm very glad to be writing about these experiments now rather than doing them.[7]

Hyphal turgor ranges from one atmosphere to 10 atmospheres, depending on the species and growth conditions. For reference, the air in a car tire is compressed to a pressure of 2 atmospheres (a little over 30 pounds per square inch). For many years, turgor was considered both a requirement for growth and an unavoidable liability for the pressurized hypha. The hypha was viewed as a clumsy contraption, something that was always close to bursting at its seams like an overstoked boiler. Because fungi insulate themselves with a resilient wall, it seemed sensible to suggest that they cannot grow without using turgor to inflate this armor. But there is a problem with the logic here, because cells without walls—animal cells—do not become pressurized. We find ourselves wrestling with a terribly circular argument. Could it be true that hyphae construct tough walls so that they can develop enough pressure to expand them? The answer to this question is clear when we recognize that the concept of turgor-driven growth suffers from a single fault: it is wrong.[8] In fact, there is no evidence at all that high levels of pressure are needed to cause wall growth. At a fundamental level, the enlargement of a walled fungal cell is no different from that of an animal cell with a more fluid surface. In both instances, water influx distends the cell, tending to give it a smooth profile, but the level of internal pressure is never a determining factor that fixes the rate of growth. Hyphae expand by allowing their wall polymers to slip past one another over their pressurized interior; they model themselves around their turgid cytoplasm. So why are fungi pressurized? When we extract these microorganisms from the artificial confines of a Petri dish and think about where they actually grow—through leaf surfaces, inside logs, between my toes—the real importance of turgor is apparent. While it isn't essential for growth when there are no barriers, pressure does enable hyphae to overcome the physical obstacles that persist after their food sources are weakened by enzymes. For this reason, pressure is a phenomenally important thing if we want to appreciate fungal biology.

If you've followed the argument about turgor pressure, you're now in a position to understand the early evolution of the organisms we call fungi. A very long time ago, the ancestors of fungi developed walls and became pressurized, adopted the form of microscopic javelins, and opened an inexhaustible menu of solid food. The formation of invasive hyphae by mushroom relatives and stramenopile fungi offers a perfect example of convergent evolution. The reason that these unrelated microbes evolved hyphae is that there are few, if any, better solutions to the challenge of penetrating solid substances.

Some of the most impressive feats of invasive growth are performed by fungi that grow deep inside nuggets of granite bedrock (Figure 3.3). Besides penetrating rock, these species also connect with the roots of trees and shrubs, forming associations called mycorrhizae ("fungus-roots"). The fungi sustain their plant partners in poor, highly acidic soils by extracting scarce inorganic nutrients like magnesium and calcium from the rocks, and acquire carbohydrates from the plants to support their own existence. This is an example of a mutualistic relationship in which neither partner can thrive without the other. Hyphae penetrate rocks by secreting compounds like citric acid and oxalic acid that

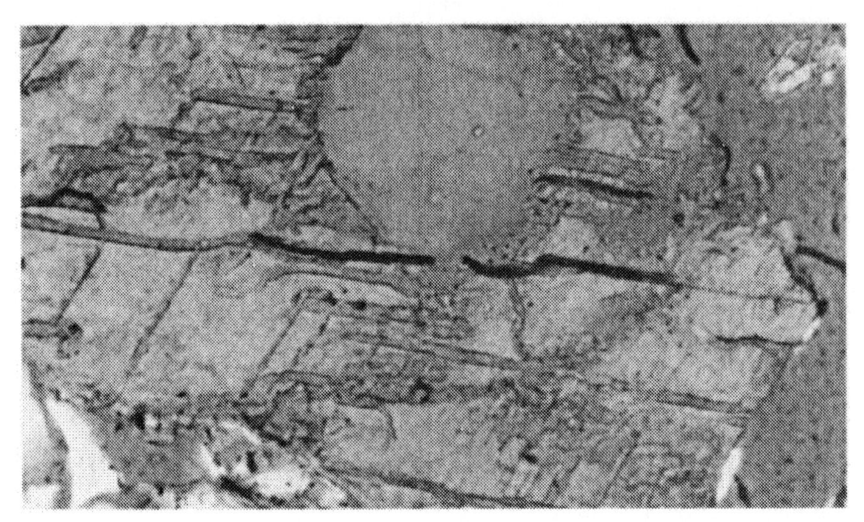

Fig. 3.3 Sample of alkali feldspar perforated with tubes formed by microscopic hyphae. The tubes with a black interior close to the center of the photograph house intact hyphae. Photograph courtesy of Toine Jongmans (Wageningen University, The Netherlands).

dissolve minerals in the granite. They absorb the leached minerals and intrude deeper and deeper, pushing forward at their tips and prying apart particles in the rock. By infiltrating granite, the fungi provide pathways for water percolation; subsequent crystallization of ice accelerates the process of weathering, which yields smaller rock fragments that are incorporated into the developing soil. Boletes and other basidiomycetes are responsible for this quarrying activity.

Rock-penetrating fungi may have been major players in the early evolution of life on land. The earliest soils on the planet were probably thin sludges derived from cyanobacteria and other simple microorganisms. These did not offer ideal conditions for luxuriant plant growth. By mining nutrients from the underlying rocks and transferring them to plants, mycorrhizal fungi may have played a crucial role in facilitating the evolution of land plants. Consistent with ancient invasive activity, fossilized hyphae of mycorrhizal fungi are found inside the root cells of primitive land plants that were preserved 400 million years ago. Far older rocks from Wisconsin contain fossils of the large spores characteristic of mycorrhizal fungi, strengthening the case for intimate relationships between plants and fungi that stretch back more than 600 million years.

The use of highly sophisticated instrumentation has allowed mycologists to measure the tiny forces exerted by single fungal hyphae. At first sight, these watery filaments do not seem particularly beefy. When the wall of a hypha yields, some of the turgor pressure inside the cell is applied against the material in contact with its apex, and the fungus pushes forward with a force of a few micronewtons (millionths of one Newton). One micronewton is produced when a mass of one ten-thousandth of a gram—equivalent to an eyelash—presses down under the influence of gravity. Not even a princess would feel it. But we are prejudiced by the fact that we generate hundreds of Newtons by sitting in a chair. This is not a fair contest. Our buttocks spread the force over a substantial contact area, so the force per unit area (or pressure) pushing down on the chair is less than a tenth of an atmosphere. Fungal forces are tiny, but because they are applied over very small areas, the corresponding pressures are surprisingly high. Most hypha exert an atmosphere or more of pressure and some pathogens that specialize in leaf penetration surge forward with 50 atmospheres or more. One of these, the rice blast fungus *Magnaporthe grisea*, was mentioned at the beginning of this chapter. Its infection cells are strong enough to pierce Kevlar®. The exploits of this fungus are featured in Chapter 9.

To evaluate the significance of the force exerted by the tip of a hypha, comparisons have been made with the mechanical resistance presented by plant tissues and other materials colonized by fungi. The strength of root tissues or a slice of agar can be measured by attaching a glass needle to a strain gauge and pushing it into the specimen. These needles can be fashioned to the same size as a hypha, allowing the investigator to attack a single plant cell under the microscope and monitor the force needed to lance its wall. Useful information can also be obtained with larger metal needles, using measurements of the force applied at the needle tip to calculate the resistance encountered by a much smaller hypha. In my laboratory we have used this approach to study the mechanical properties of skin samples from horses and humans. Fresh samples of horse skin can be obtained with relative ease, but the procurement of human hide presents a much greater challenge. The first time I attended an autopsy, I was alarmed to find myself in a room with dead people. What had I expected? But after a few seconds, the sense of surprise was overwhelmed by my fascination with the corpses and trays of surgical instruments, and an eagerness for fresh skin samples. The first patient had died a few hours before I saw her. The woman's head was propped on a bloody wooden yoke and her tongue had swollen out between her lips. Once the butchery began, I hovered between the pathologist and medical residents like a vulture, holding my ice box in readiness for the skin strips. For a few weeks after my work with morsels of human skin, my passion for sushi was severely challenged. The thin cutaneous layer, with its underbelly of yellow fat, bears an awful resemblance to fresh mackerel and other Japanese delicacies. Horse skin causes no problems. After all, few fish sport such a luxuriance of coarse hair.

The skin of humans and horses offers about the same resistance to needle insertion, averaging between 200 atmospheres and 300 atmospheres of pressure (equivalent to a mass of a few kilograms acting on the tip of a needle). Since most hyphae exert pressures of one atmosphere or so at their tips, there is no way that these microorganisms can grow through undamaged skin without using enzymes to soften this proteinaceous tissue.[9] Wounds offer microbes an alternative to piercing the skin barrier for themselves, which is why surgery, bullets and shrapnel, barbed wire, and even rose thorns extend invitations for fungal colonization. Strength measurements from roots and leaves suggest that, like human pathogens, the fungi which infect plants partially dissolve the polymers that comprise the solid tissues of their prey during invasion. Irrespective of the composition of the substrate, the fungus must be braced in some fashion to avoid ejecting itself as it thrusts against the end of its microscopic tunnel. (This is a simple consequence of Newton's third law of motion regarding equal and opposite reactions.) So in the earliest phase of colonization, spores that land on the surface of plant and animal tissues secrete potent adhesives to provide firm attachment before the formation of invasive hyphae. Later, when the fungus is buried inside the substrate, friction between the convoluted surface of the growing mycelium and the enveloping material supports the extending tips.

When fungi have exhausted their food sources they play the invasive growth video in reverse, bursting into the air and forming spores. If the substrate for growth has been highly degraded, little of the mechanical barrier that impeded invasion remains, and the fungus can probably reemerge without further secretion of enzymes. But even in this situation, the water-saturated environments occupied by fungi present a further obstruction in the form of surface tension. If the mycelium is covered with a film of water, the hyphae must overcome the elasticity of the interface between water and air and this can be a significant hindrance. The force necessary for a hypha to stretch and break a film of water is in the micronewton range, matching the force brought to bear during invasive growth. Some species reduce the surface tension of water by creating the counterpart of an oil slick. They do this by secreting hydrophobic proteins that congregate into a thin layer at the air-water interface, and this slippery film aids hyphal emergence in preparation for spore production. While overcoming surface tension can be challenging for individual hyphae, by collaborating to form a mushroom, the protrusive activity of millions of cells—each one exerting a few micronewtons of force—is sufficiently powerful to crack a paving slab.

While mycologists have succeeded in demystifying the mechanics of hyphal growth, despite many years of inquiry we remain stumped by the mechanisms that allow these cells to produce their cylindrical shapes from smooth-domed tips. This sculpting process is known as morphogenesis.[10] The polarized delivery of vesicles to the hyphal tip is central to the operation, and the details of this transport mechanism are becoming clear. But what informs the cell about the placement of its tip? Various molecules seem to mark the hyphal apex, but what establishes the position of these labels? Mycologists are not alone in their confusion. Morphogenesis is a puzzle of encyclopedic proportions for all biologists. Some of the leading proponents of genomic analysis are optimistic that a full reading of the genes of a particular type of cell (or multicellular organism) will indicate how shaping takes place, but their self-promotion has gone too far here and few believe them. Part of the problem is that genes specify proteins, not cell shapes, at least not directly. Even a complete catalog of the proteins is insufficient, because shape materializes from interactions among vast numbers of proteins, other chemical constituents, and the environment occupied by the cell. The complexity of the living thing is unveiled through development as layer upon layer of structure and behavior build on the preexisting organism. Zygote becomes embryo becomes baby becomes adult; spore becomes hypha becomes mycelium becomes mushroom. Even a partial solution to this enigma would allow mycologists to come much closer to understanding how a single spore produces a soil-bound leviathan, fells a tree, or ends your life.

CHAPTER 4

Metamorphosis

When Gregor Samsa awoke one morning from uneasy dreams he found himself transformed in his bed into a gigantic insect.
—Franz Kafka, *The Metamorphosis* (1915)

Faced with a classroom of sleepy students, an English professor wielding the novels of D. H. Lawrence usually enjoys a more gratifying response than the Ph.D. pointing to a diagram of a mushroom gill. Mycology can be a tough sell. Every fall I lead a ramshackle group of unfortunate young men fatigued by years of video-gaming in dark basements and their indifferent female counterparts into the woods. In wet years, when the mycological carnival is in full swing, the fungi and I manage to bewitch one or two in my flock. Moss-blanketed logs that slump beside a creek on my university campus aid in this seduction. The students have no idea what I'm doing when I drop to my knees and begin to scan the velvety surface of one of these decomposing trees; I'm looking for swollen-tipped aerials that spring upright between the whiskers of the moss. Digging into the damp wood with a pocket knife, carefully pursuing the root of an aerial, I expose a dead insect, sometimes an adult beetle or a larva, more often a wood ant, wrapped in a shroud of hyphae. The inch-high stalk that sprouts from the corpse is the fruiting body of a *Cordyceps*, a platform that broadcasts infectious spores from its tip (Figure 4.1). In the arena of mycological dramatics, nothing is more electrifying than the act of extracting a mummified insect from the heart of a rotting log.

If you can muster any empathy toward an insect or spider, *Cordyceps* can seem a terrifying adversary. The fungus is capable of piercing the

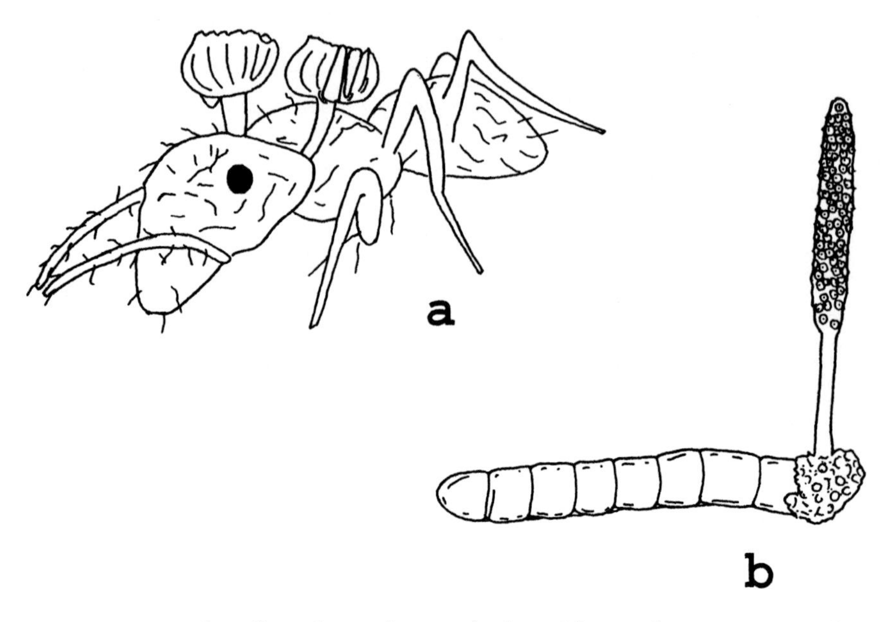

Fig. 4.1 Examples of *Cordyceps* fruiting bodies. (a) Paired ascocarps growing from the neck of an ant. (b) Single ascocarp with pimpled tip attached to mummified insect larva.

chitinous exoskeleton of insects; once it is inside, the symptoms of disease become obvious. Ants colonized by a *Cordyceps* engage in hopeless grooming behavior and move with an erratic gait, their legs twitching uncontrollably. In the most spectacular displays of torment, epileptic insects wracked with convulsions climb plant stems and ooze mucus from their mouthparts. In their death throes they grip the top of the stalk with their legs, clamp down with their can-opener mouthparts, and expire. Sometimes, the bodies of the dead insects are scavenged by other ant species and all that remains is the infected head secured to the stalk by its powerful jaws. Other insects show the reverse behavior following infection, moving down a tree trunk or burrowing into wood (like the ones I find in Ohio). But for the pathogen, the outcome never varies. A few days after death, its hyphae burst through joints in the insect's skeleton, cloaking the animal in mycelium before sprouting the elongated fruiting body called the ascocarp (or "stromatic clava," according to some specialists). Sometimes, the ascocarp erupts from the head of the insect. In his book *Mr. Wilson's Cabinet of Wonder*, Lawrence Weschler[1] describes his astonishment at reading a description of a *Cordyceps* sprouting from a Cameroonian stink ant. He couldn't believe the bizarre story of the tormented insect's climbing response until he telephoned a biologist who told him this was true.

The climbing behavior is also observed in insects infected with other kinds of fungi. The term *summit disease* is applied to insects manifesting this symptom. By scaling vegetation, colonial insects remove themselves from their nests, limiting the opportunity for an epidemic. The self-burial strategy may also have evolved to protect the sickened insect's relatives from infection. Such gallant and seemingly altruistic actions make sense for sterile animals like worker ants, which may have hundreds of thousands, if not millions of identical sisters. In biological terms, selfless actions serve oneself in the colony because the genetic legacy of each worker depends entirely upon the health of the queen and her suitors. Alternatively, the fungus might be exerting direct control over the insect's behavior by manipulating its brain chemistry. Although purely hypothetical, the neurological symptoms exhibited by infected insects make this a very provocative idea. By inducing the summit response in its host, the fungus would place itself in an ideal position for wind dispersal of its spores, provided that once exposed, the insects are not picked off by birds.

Interactions between fungi and insects often defy a simple explanation. My friend Nigel Hywel-Jones (the gentleman with athlete's foot in the second chapter) is an expert on these pathogens; he has spent fifteen years searching for new species in the rain forests of Thailand. He's found *Cordyceps* on cicadas, tarantula spiders, ants, and termites. Termites infected by one of Nigel's recent discoveries die in pairs, harnessed to one another's jaws by a weft of fungal mycelium. Perhaps one termite is infected by the other when it attempts to groom its diseased kin.

Cordyceps has been used in Chinese medicine for hundreds of years and is now marketed by American health food stores as a cure for every conceivable ailment. In China its fruiting bodies are called dong chong xia cao, which means "winter insect, summer grass," evoking the apparent conversion of caterpillars into roots from which the ascocarps (which do look a bit like plant stems) extend into the air (Figure 4.1 b). The colonized insect is sold in a dried form that resembles a twig. Usually, dong chong xia cao is powdered and made into a tea or mixed with other ingredients and swallowed as a tablet. Thanks to the Internet, this Oriental medicine has become widely available, offering disease-free lives for all humankind. According to one site, *Cordyceps* assists body building, overcomes fatigue, improves lung and kidney function, and—least surprising of all—increases male sexual potency. There is also interest in its efficacy as a performance-enhancing drug for marathon runners. Although I have tried to dissuade everyone from consuming phallic mushrooms, I threw caution to the wind with this Chinese elixir and ordered some *Cordyceps* tea from a company called Fungi Perfecti, based in Washington State. (The president of the company, Paul Stamets, is an expert on hallucinogenic mushrooms who has developed a successful business that markets dietary supplements and medicines prepared from fungi, and spawn for growing edible mushrooms.) Unfortunately, I could not stick with the *Cordyceps* brew for the three weeks recommended for optimum manliness. Although I have downed more frightful beverages, this one is no Earl Grey.

Cordyceps is an example of an ascomycete, the largest grouping of fungi that encompasses more than 32,000 species that occupy every ecosystem on the planet. They are defined by the production of a unique kind of spore called the ascospore, which is formed inside a fluid-filled purse called the ascus (Figure 4.2). As cytoplasm within this cell is reorganized, embryonic spores are delineated by a pair of encircling membranes, and a clear juice appears between the spores and the wall of the ascus. Water diffuses into the ascus by osmosis, and the structure becomes pressurized like a hypha. When the tip of the ascus opens, often in an explosive fashion, the spores are shot into the air. This discharge mechanism is best described as a cannon and bears no relationship to the surface tension catapult utilized by basidiomycetes.

My local *Cordyceps* is dwarfed by tropical and subtropical species with large, brightly colored fruiting bodies. The ascocarp of an Australian fungus called *Cordyceps gunnii* is as fat as a garden hose and longer than a stick of celery. This species infects a root-feeding caterpillar and then elongates until its tip emerges into the air through the opening to the insect's burrow. The upper portion of mature *Cordyceps* ascocarps is covered with pimples that mark the openings of tiny flask-shaped chambers called perithecia, in which the asci develop. A single perithecium holds many asci, and each ascus contains eight ascospores (Figure 4.2 a). When each ascus matures, it elongates, pushes its tip through an opening in the center of the pimple, perforates at its apex, and fires its clutch of thread-shaped spores, one at a time, before retracting into the perithecium. A few minutes later, the next ascus protrudes from the pimple and discharges its spores. Each spore spans half a millimeter in length and is built from 100 or more segments that fragment in the air or when they hit something. This murderous projectile is the ultimate fungal weapon, a missile with numerous warheads. Because the number of warheads or spore segments is multiplied by the number of ascospores, and then by the number of asci per perithecium, and the number of perithecia, it is clear that a single fruiting body can dispense millions of infectious particles. This staggering disease-causing potential is the result of an arms race between *Cordyceps* and insects that has probably been waged for hundreds of millions of years.[2] For example, it is likely that the evolution of the monstrous *Cordyceps gunii* has involved the progressive elongation of the fruiting-body stalk so that this pathogen is able to escape the deepest graves of its victims. Conflict at the biochemical level has probably been equally fierce, with the proliferation of toxic and antitoxic cocktails of ever-increasing strength within both host and pathogen.

Cordyceps poses no threat to humans, but its relative *Claviceps* is a source of great misery. *Claviceps* is the ergot fungus, a pathogen of rye whose toxins cause blood vessels to constrict so powerfully that hands and feet become gangrenous, causing patients to shed their nails and, eventually, their hands or feet. I know this sounds like a sketch from *Monty Python's Flying Circus*, but a report of ergot poisonings in England in the eighteenth century[3] referred to "a singular calamity, which suddenly happened to a poor family in this parish, of which six persons lost their feet by a mortification not to be accounted for." The vasoconstricting poison, called ergotamine, is preserved during baking, and most poisoning cases have resulted from the consumption of contaminated rye bread, which served as a staple food in parts of Europe for centuries and in colonial America.[4] Initially, patients feel agonizing burning sensations in the arms and legs, and in some cases the pain is accompanied by terrifying hallucinations. The hallucinations are caused by another chemical synthesized by the ergot fungus: isoergine or lysergic acid amide, a less potent version of lysergic acid diethylamide, or LSD. The synthesis of isoergine is a nice touch by the fungus, an additional torment for the afflicted (I doubt that anyone in the Middle Ages suffering from gangrene and the belief that they were being attacked by demons enjoyed an unintentional acid trip).

In the 1970s, Linnda Caporael, of the Rensselaer Polytechnic Institute in New York State, proposed that the demonic possession ascribed to women in Salem, Massachusetts, in the seventeenth century was caused by ergot poisoning. The women suffered epileptic convulsions and choking, and reported feeling as if they were being pinched, bitten, and pricked with pins. The spring and summer of 1691 were unusually warm and wet, providing perfect conditions for *Claviceps*. Symptoms of poisoning began in December, immediately after the threshing of Salem's grain harvest, some of which was almost certainly contaminated with ergot. The devil's blight persisted for a year, finally disappearing quite abruptly, following the drought of 1692. *Claviceps* doesn't thrive in dry years. In all likelihood, the women recovered once they began eating bread made from clean flour. Whether ergotism explains the whole tragedy, any rational explanation is three centuries too late for the twenty colonists executed for witchcraft.

The ascomycetes are so diverse in their form and lifestyles that they deserve their own book, but I'll illustrate this variety with a few examples. It is impossible to talk about ascomycetes without mentioning the most famous of all fungi: yeast. A packet of freeze-dried yeast costs just a few cents but contains a marvel. A few teaspoons of water transform this powder into life, creating a sludge that will make bread dough rise, turn sugar to alcohol in beer and wine, and facilitate human existence over much of the planet. In historical terms, our relationship with yeast is more intimate than any other between fungi and humans.

Yeast is to fungi as whale is to mammals. There are lots of species of yeasts and whales. Nevertheless, "yeast" is often used to denote a single species, baker's yeast, or *Saccharomyces cerevisiae*. This reflects the importance of *Saccharomyces* to human nutrition and to biological research, but hundreds of fungi have a similar growth form. The cells of *Saccharomyces* are usually egg-shaped, although they sometimes expand as perfect

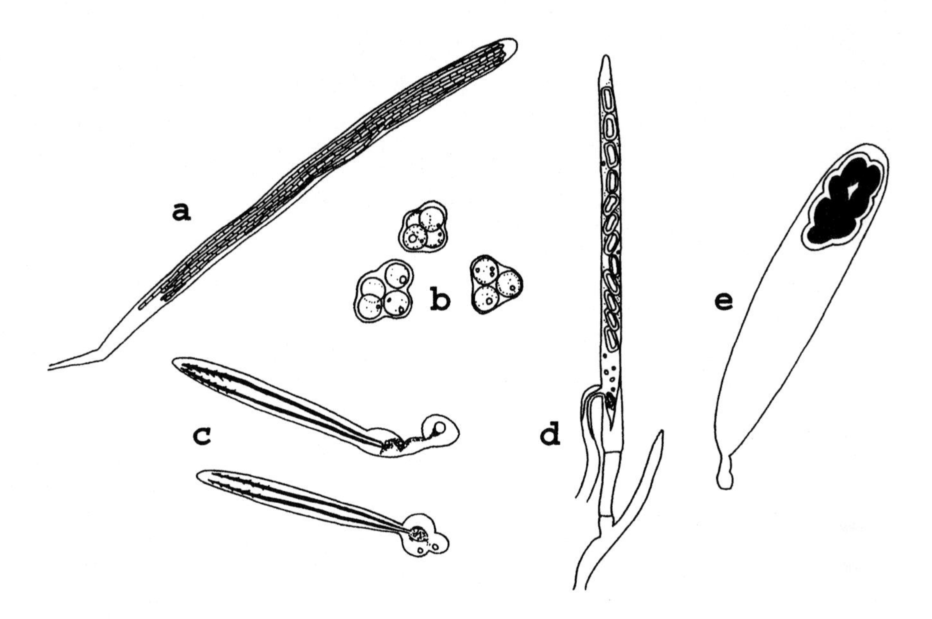

Fig. 4.2 Asci. (a) The ascus formed by *Cordyceps* species is a quiver that contains eight elongated spores that fragment into multiple infectious cells. (b) *Saccharomyces cerevisiae*. (c) The predacious yeast *Metschnikowia hibisci*. Each ascus contains two needle-shaped ascospores with barbed tips. (d) The individual spores of *Dipodascus macrosporus* are surrounded mucilaginous sheaths that lubricate the spores as they are ejected from the ascus. (e) The ascospores of *Ascobolus immersus* are packaged in a common mucilaginous coat so that they are ejected as a group. Not drawn to the same scale.

spheres. They reproduce without sex, by dividing the nucleus and moving one of the resulting daughter nuclei into a bud or daughter cell that swells from one or other end of the mother. In this way, a family of daughters cracks from the mother's surface to join the ever-multiplying fray, each daughter leaving her parent with a circular bud scar in the cell wall. The surface of the daughter is also scarred by the fungal counterpart of a navel (called the birth scar), which marks the point of separation from the mother cell. Daughters become mothers, and with no need for a mate they thrive in a manner that would gladden the most militant feminist. But yeasts have sex also. There are two strains of *Saccharomyces*, called alpha and beta, that signal to one another using chemicals called pheromones, make contact, fuse, and finally, develop ascospores inside the shared wall.

The whole organism, which exists as a single cell, is converted into an ascus (Figure 4.2 b). Unlike *Cordyceps*, *Saccharomyces* doesn't expel its spores with a cannon but spills them from the ascus when its wall dissolves. Each of the spores germinates to produce an alpha or beta yeast. Male and female designations are without meaning here.

To understand the value of the yeast form, it is useful to think about nutrient uptake. Yeasts and hyphae release digestive enzymes and absorb small nutrient molecules through their cell walls and membranes. In the previous chapter, I explained that the structure of a hypha makes sense when it is viewed as a device for mining food from materials like decaying wood: the tip of the cell forges through solid substances, creating a cylinder of ever-increasing length. Because most of the nutritional needs of a mycelium are met by food absorption through the apices of its hyphae, the older core of the organism dies as the fungus widens its territory. For mechanical reasons, the yeast is not an efficient vehicle for movement through decaying logs or soil, but in fluid environments, this compact, single-celled form is far superior. The yeast absorbs food over its entire surface, providing the cytoplasm with a rich supply of nutrients far more easily than a hypha, whose feeding activity is confined to a tiny fraction of its exterior. Many fungi switch between the yeast and hyphal conditions as dictated by the physical and chemical nature of their surroundings. Even *Saccharomyces* constructs a rather pitiful kind of hypha (a lumpy chain of yeasts) when it is forced to do so by nitrogen limitation, and is then able to penetrate the agar in its culture dish.

We know more about the composition of *Saccharomyces* and how it works than any other organism with a nucleus. Experiments on this species have furnished lessons on the way that every living thing functions, and in 1996, biologists divulged the sequence every one of its 6,000 or so genes, marking the publication of the first eukaryote genome. But hard as I have tried, I've never felt excited by this simplest of fungi. Only a biochemist can be truly satisfied by such dull architecture when the grandeur of other fungi can surpass the Palace of Versailles. This is evidenced at scientific meetings: *Saccharomyces* shows some scars and a bud and requires a vibrant speaker to champion its graces; photographs of the fruiting bodies of some *Cordyceps* species, presented by a gentleman who mumbles into his beard, are sufficiently lovely to make an audience gasp with delight. I'm expressing prejudice here of course, and recognize that we have learned far more about biology from studying yeast than ascomycetes that attack insects. Yeast experts are similarly intolerant. For decades they have objected to the appellation of mycologist, fearing that their reputations as serious laboratory scientists will be undermined by popular images of people who collect mushrooms. Such concerns may be well founded for scientists who obtain funding for yeast research by selling *Saccharomyces* as a flawless model for understanding human cancer. In any case, *Saccharomyces* is a tiresome thing to look at.

Unlike *Saccharomyces*, there are other yeasts whose shape becomes greatly modified when they produce spores. For example, species of predacious yeasts attack insects and crustaceans like brine shrimp, by lancing them with needle-shaped ascospores. Some of these live in the flowers of *Hibiscus* and morning glory that open for just one or two days before closing and embarking on seed production. The asci of these yeasts are elongated spore guns that discharge pairs of spores decorated with spirally arranged barbs that point away from the tip (Figure 4.2 c). Although the act of assault has not been witnessed, these spores seem perfectly suited for harpooning beetles that visit the flowers.

As a student I worked with another yeast, called *Dipodascus*, which grows in the mucus trails left in the wake of slime molds. Removed from slime mold "feces," and spread on an agar plate, the yeast grows into a mycelium of branched hyphae, rather like the trick performed by the human pathogen *Wangiella* (Chapter 2). The conspicuous feature of *Dipodascus* is its upright, tapered asci that fill with ascospores (Figure 4.2 d). Each spore is wrapped in a thick, mucilaginous coat. When the tip of the ascus ruptures, the heavily lubricated spores slide out and form a loose cluster. Looking down on a *Dipodascus* colony, one sees a dew-saturated raft of ascospores held aloft on the asci and aerial hyphae. It's a little more appealing to look at than *Saccharomyces*, but still lacks a multicellular fruiting body.

Relatively simple ascomycete fruiting bodies (or ascocarps) are closed spheres called cleistothecia that contain one or more asci. *Eurotium* is one of these cleistothecial ascomycetes (Figure 4.3). Its minute fruiting bodies are built from a jacket of hyphae that become yellowed with age,

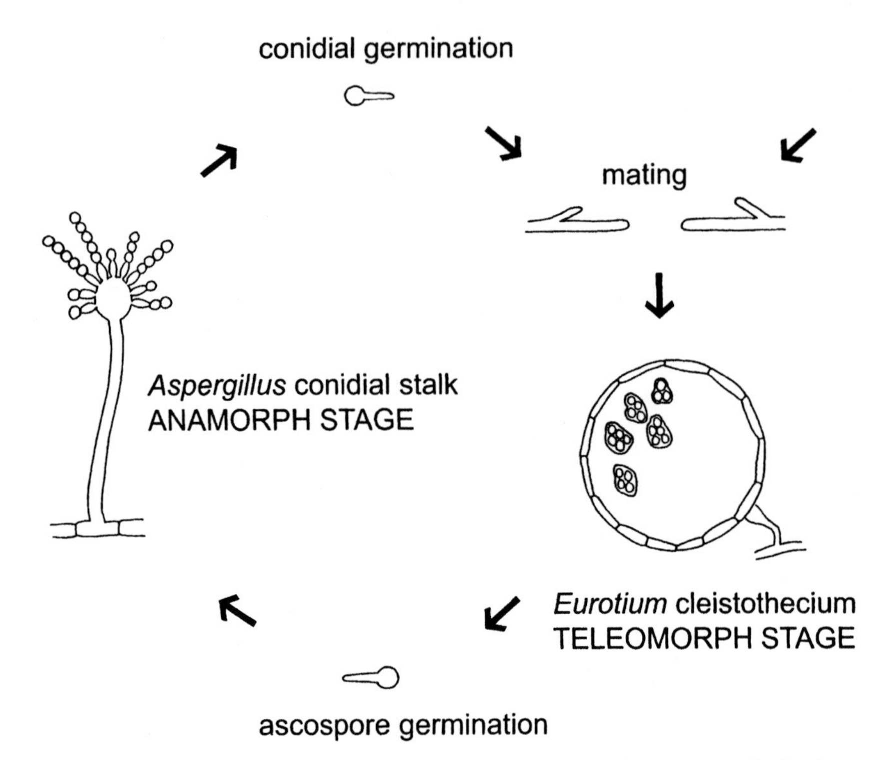

Fig. 4.3 Alternation of asexual (anamorphic) and sexual (teleomorphic) phases in the life cycle of *Eurotium/Aspergillus*.

and the asci that develop inside are flaccid bags that dissolve like those of the yeast, turning the cleistothecium into a spore-filled globe. Eventually, the wall of the ascocarp breaks open, freeing the spores. But the cleistothecium is only half the story. Like most ascomycetes, *Eurotium* leads a different life in the absence of a sexual partner. A single ascospore of *Eurotium* will germinate and produce a branched mycelium that invades agar medium in the laboratory and spoils groceries in the kitchen. At the surface of its food source, the fungus sends branches into the air that bulge at their tips and generate massive numbers of spores. This process of abundant spore formation is achieved without any mating, which means that the nucleus inside each spore is a clone of the nuclei in the parent. These asexual spores are called conidia.

Here's where things get tricky. When *Eurotium* forms conidia, mycologists refer to the fungus by a different name: *Aspergillus*. This conidial stage is the imperfect state or anamorph of *Eurotium*; *Eurotium* is the perfect state or teleomorph of the same organism. If a pair of *Aspergillus* colonies with compatible genetics merge with one another, they will produce the cleistothecium of *Eurotium*. *Eurotium* forms *Aspergillus* forms *Eurotium*, in an endless cycle of sex followed by celibacy followed by sex (see Figure 4.3). We encountered the same phenomenon with the human pathogen *Cryptococcus* and its sexual, basidiospore-producing *Filobasidiella* stage. Again, the reason a single organism is given two names is due to an incomplete understanding of its life cycle.

It is a simple matter to fish an *Aspergillus* from the sky. Open a Petri dish anywhere outside the sterile air of an operating room for a few seconds, close the lid, and leave it for a day or so, and the plate will likely sprout the conidium-producing stalks of an *Aspergillus*. If the experiment fails, the fungus is certain to appear on the second or third attempt. Once isolated, the culture can be maintained in the laboratory indefinitely by seeding fresh agar medium with spores from an older plate. Mycologists who spend their professional lives studying and naming species are responsible for determining whether particular fungi have been described by someone else, or are new to science. Imagine that we have discovered a new species of *Aspergillus* growing on a culture plate. We'll name this fictitious organism *Aspergillus magnificus*. If two strains of *Aspergillus magnificus* are grown on the same agar, they will mate and generate a sexual fruiting body. Since we're dealing with an *Aspergillus*, this fruiting body will be a cleistothecium. A mycologist who comes across the cleistothecia formed by two strains of *Aspergillus magnificus* on the surface of a leaf might very well think that he or she has discovered another new species. Unless the conidium-producing *Aspergillus* stage in the life cycle is growing alongside the cleistothecia, then the error is unavoidable, and the fungus producing the cleistothecia will be given a separate name: *Eurotium grandiosum* sounds good. For this reason, a vast number of fungi have been described twice because the observer was privy only to one chunk of a larger life cycle.

This confusing predicament—and there is nothing more perplexing in the entire field of mycology—was first recognized in the middle of the nineteenth century by two bachelor brothers named Louis-René and Charles Tulasne. The brothers' contention, that individual fungi could produce different types of spore, was made, with frequent references to the glory of God (they were devout Catholics), in an enchanting three-volume tome entitled *Selecta Fungorum Carpologia*.[5] This principle was termed, variously, multiple gemmation, polymorphism, and pleomorphy, and its authors revolutionized the study of mycology. In their younger years Louis-René had studied law and Charles became a physician, but later in life, aided by an inheritance from their father, they dedicated themselves to mycological studies. Louis-René had acquired much of his scientific knowledge after he quit the legal profession and pursued a research career at the Jardin de Plantes in Paris. He wrote most of the text of the *Carpologia*, while his brother drew the luminous illustrations for each massive folio volume. Charles created awe-inspiring three-dimensional depictions of fungi whose information content far exceeds that of any photograph.

This superiority of the best drawings or paintings over a photograph is rooted in mechanics of these artworks. Freed from the search for a single flawless composition for a photograph, the illustrator can incorporate his or her impressions of multiple individual specimens into the portrait of a particular organism, without having to contend with shadows that obscure parts of the structure in a photograph. With the same mastery that Audubon (and, more recently, Sibley)[6] profiled American bird species for serious ornithologists, Charles Tulasne offered caricatures of fungi that have dazzled mycologists ever since. It is as if he shrank himself to the size of a spore and drew the fungal forest that towered around him. The drawings allow one to peer at the fluid interior of cells through their transparent walls, sense the roughness of spore surfaces, and imagine the tactile pleasure of running a finger over the roughened exterior of a fruiting body.

Initially, the concept of pleomorphy met significant opposition from contemporaries of the Tulasne brothers, and then, with similar fervor became universally accepted, and then misapplied to the point of lunacy. Ernst Hallier, professor of botany in Jena, Germany, suggested that the yeast *Saccharomyces* turned into a specialized insect pathogen when it was consumed by flies, and that it could give rise to a bread mold in air, or a zoospore-producing oomycete if the insect fell into water. In short, Hallier was able to subsume a ream of unrelated fungi into a single species by suggesting that this flexible microbe possessed a life cycle that—in my opinion—would rival the escapades of a worm that was capable of modifying itself into an elephant or a sea snake. Even more absurd, Hallier became convinced that the bacterium that causes gonorrhea was actually a stage in the life cycle of a fungus that produced conidia when it was enjoying life outside someone's genital plumbing. Although ridiculous, it is useful to consider that the theory of spontaneous generation—which in its earliest rendition held that rodents could originate spontaneously from piles of garbage[7]—was not disproved until the publication of Louis Pasteur's experiments with "swan-necked" flasks in the 1860s. In his landmark book on fungi published in English in 1887,[8] Anton de Bary discussed the work of the Tulasne brothers and accused Hallier of "pleomorphic extravagances." Hallier raised no defense against this public defrocking, and instead allowed himself free rein to ignore rational thought entirely by abandoning science and devoting his genius to the study of philosophy and aesthetics.[9]

The concept of pleomorphy is particularly fascinating from a genetic point of view. An individual fungal genome contains all of the information necessary to supervise a profound reorganization of an organism whose sexual and asexual phases often represent radically different solutions to survival. One expression of the genome may generate spores in air, the other in water; one thrive on a leaf, the other in a human lung, and in each location the fungus looks and behaves like a distinct species. The *Aspergillus* stage of *Eurotium* looks nothing like *Eurotium*, and everyone in the field is eager to learn about the changes in gene expression that result in this metamorphosis. We know already that the process of modification may be rather subtle, involving far fewer changes at the molecular level than one might have predicted. There are no great swaths of DNA that contain the specific instructions for making conidia rather than ascospores, but instead, as the fungus begins to produce its asci, flashes of activity occur in regions of the genome (genes or suites of genes) that until then were quiet, and other genes are silenced. Understanding how these alterations result in the production of a multicellular fruiting body like a cleistothecium is the next great challenge for mycology. Related questions are invigorating the whole field of developmental biology, and whether the inquiry is directed at fungi or apes seems of secondary importance. Personally, I find Charles Tulasne's drawings of ascomycete development far more remarkable than the images in my mother's photograph album. While I suspect that my initial metamorphosis from egg to embryo was thrilling, a glance in the mirror shows that after forty years I'm still wearing the silly grin I concocted as a gurgling infant, and my arms and legs sprout from the same general locations. Even after Kafka turns his protagonist into an insect, the reader never loses sight of Gregor Samsa. The identity of a fungus is almost always in question.

Let's look at the biology of *Aspergillus* in more detail. Some species of *Aspergillus* synthesize cancer-causing compounds called aflatoxins in foods (I'll discuss these in a later chapter), and others can cause allergies, colonize the lungs, and initiate invasive infections in patients with weakened immune defenses. On a brighter note, soy sauce is produced when a pair of *Aspergillus* species are allowed to ferment cooked soybeans. In common with many of the conidial stages of other ascomycete fungi, *Aspergillus* species are described as generalists because they will grow and reproduce on an assortment of substrates. In the solitude of their conidial phases, these fungi do not produce spores inside asci, but shape them from the tips of modified hyphae. As I've said before, almost everything fungal has a hyphal origin. The spore-producing stalk of an *Aspergillus* is a hypha that emerges from its food source and grows into the air. Once the stalk has elongated, its tip swells and an array of specialized cells called phialides arise on the rounded head. Phialides are shaped like bowling pins. The nucleus inside the phialide divides continuously, and with each duplication, one of the daughter nuclei moves to the tip of the cell, where it is packaged into a spore and extruded like a blob of toothpaste. This process is repeated every few minutes so that each philade gives rise to a chain of spherical conidia. The fact that *Aspergillus* strains can be isolated anywhere on the planet is a testament to the effectiveness of this asexual reproductive mechanism.

Penicillium species, famous for their gift of antibiotics, manufacture chains of spores from phialides like those of *Aspergillus*, but there are countless alternative mechanisms for clonal replication among conidial fungi. Conidia can be produced by the fragmentation of a mycelium at its septa, or from yeast-like branches that separate from the parent mycelium; conidia of other fungi inflate from pegs on the surface of aerial hyphae, and still others create conidia that produce more conidia themselves, cultivating ever-extending chains of spores. The details of these processes fill the pages of mycological journals. A few minutes spent flipping through *Mycotaxon,* a periodical for fungal taxonomists, offers a window into a bewildering display of fungal structures. Complexity escalates because the conidia themselves are tremendously varied in shape and size. The spheres of *Aspergillus* are fairly simple, while other conidia are elongated, multicellular, branched, coiled, *S*-shaped, and even star-shaped. Finally, to make sleep impossible for the student of fungal identification, many species cluster their conidium-producing hyphae into bundles, tufts, and even multicellular constructions that are as complex as the fruiting bodies of fungi that produce sexual spores.

To make sense of this bestiary, taxonomists have advocated hundreds of terms to describe the shape and developmental origin of spores. The 9th edition of the *Dictionary of Fungi*[10] defines more than 120 of these, including "dictyochlamydospore" and "botryo-aleuriospore." For anyone other than a specialist in fungal identification, the terminology is dumbfounding, and even among experts, disputes about the application of particular terms complicate the description of newly discovered fungi. Carolus Linnaeus, who invented the binomial system for naming organisms in the eighteenth century and managed to describe more than 12,000 species of plants and animals himself, was perplexed by the variety and plasticity of the fungi and banished a diverse selection to a species he named *Chaos fungorum*.

At an elementary level, we can recognize three categories of ascomycete fruiting body: the closed fruiting bodies exemplified by *Eurotium* called cleistothecia; the flask-shaped perithecia of *Cordyceps*; and the cup fungi, whose asci are exposed at the surface of open ascocarps called apothecia (Figure 4.4). The apothecia of *Ascobolus immersus* are unassuming, gooey yellow blobs that grow on herbivore dung, but the asci are astonishing: a succession of these crystal-clear cannons extends from the blob to blast their octets of purple spores into the sky (Figure 4.2 e). The ascospores are sheathed in mucilage, which holds them together during flight, and the large projectile lands up to 30 centimeters from the launch pad. By expelling its spores as a single, larger mass, *Ascobolus* maximizes the range of its asci, because the motion of

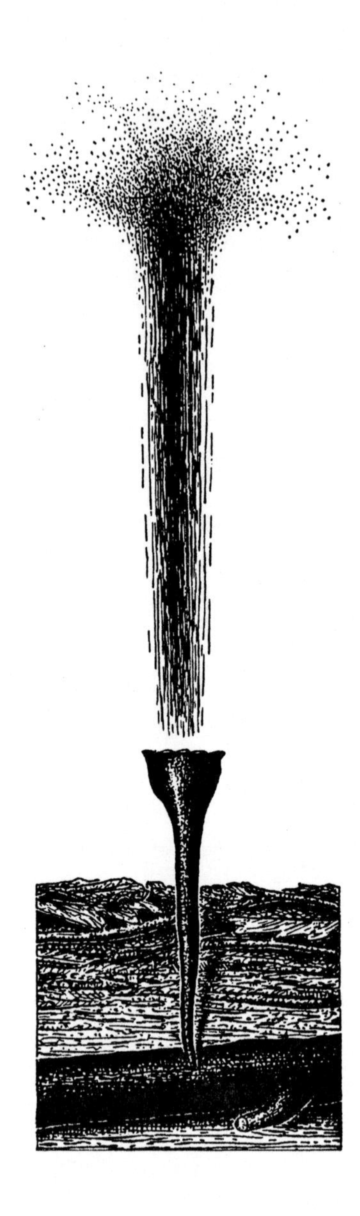

Fig. 4.4 Apothecium of the cup fungus *Microstoma protracta* discharging masses of ascospores by the simultaneous discharge of numerous asci. From A.H.R. Buller, *Researches on Fungi*, vol. 6 (London: Longmans, Green, 1934). According to Buller, puffing of spores can be heard as a hiss by holding an apothecium close to one's ear. By inserting its open end inside the ear, the blast of spores is felt "as though the ear were being sprayed by a fine atomiser."

the separated spores would be severely limited by air resistance (consider that the surface tension catapult mechanism propels each mushroom spore for no more than a few tenths of a millimeter). The necessary force is provided by a pressure of about 2.5 atmospheres inside the ascus, which is relieved when a lid at the tip of the cannon flaps open at the moment of discharge. Each ascus orients itself toward the sunlight as it extends above the glistening surface of the fruiting body. By automatically bypassing any flight path shadowed by overhanging vegetation, this so-called phototropic behavior maximizes the chance that once they are discharged, the spores will be projected as far as possible from their launch pads. Like the bird's nest fungi described in the first chapter, *Ascobolus* spores are eaten by animals that will graze close, but not too close, to their own dung.

Ascobolus immersus is endowed with the largest ascus of any fungus—a tenth of one millimeter in width and a millimeter in length (see Figure 4.2 e)—but the whole fruiting body is just one or 2 millimeters in diameter. Like *Ascobolus*, most ascomycete fruiting bodies are much smaller than the mushrooms of basidiomycete fungi, but there is, or once was, an extravagant exception.[11] A South American species called *Geopyxis cacabus* is alleged to construct a one-meter-tall stalk that supports a spore-producing cup as wide as the lid of a trash can! Packed with asci, such a cup could launch as many as 7 billion spores.[12] As impressive as this sounds, this is a thousandfold less than the estimated output of the most fecund organism on the planet: 7 trillion spores can smolder from a single giant puffball.[13]

Aside from their sheer numbers, another characteristic of ascospores is their resilient nature. Nick Read, a mycologist at the University of Edinburgh, published a fascinating study in which he demonstrated that ascospores can survive the brutal environment within an electron microscope and germinate after removal.[14] In Nick's experiments, the spores were subjected to temperatures as low as –180°C, dried in a vacuum, and then irradiated with an intense beam of electrons in the microscope. These experiments lend credence to the claim that in addition to the apparently indestructible body of Keith Richards, certain microbes might survive apocalyptic changes in the Earth's climate resulting from ozone thinning, global warming, or nuclear war. There may be a further lesson in Nick's research. He discovered that only one kind of abuse killed every ascospore: dehydration in alcohol. In their dried state, lichens are also phenomenally hardy. Again, for the purpose of electron microscopy, a common species was bathed in liquid nitrogen, fractured with a steel knife, and coated with gold and palladium. Following these insults, the lichen was glued to a ceramic specimen holder. It grew normally, extending its thin fronds into the air as if nothing untoward had transpired.

Lichens are fungal mycelia whose exposure to air and sunlight is made possible by a combination of extraordinary physiological adaptations coupled with their partnership with other microorganisms. Apothecial fungi are the most common fungi in lichen associations with photosynthetic algae or cyanobacteria. Cells of the alga or bacterium occupy a discrete layer just beneath the lichen's surface, where they receive sufficient light for photosynthesis but are shielded by the fungus from damage by ultraviolet rays. Ultraviolet light, which is far more disruptive than the radiation inside an electron microscope, is absorbed by melanin within the cell walls of the hyphae and causes the lichen to warm. In this way, the pigment may delay freezing of the lichen and prolong metabolic activity at low air temperatures. As an undergraduate I studied lichens on granite boulders in Snowdonia National Park in Wales. Brightly colored crusts covered the most richly fertilized rocks, and the ravens that encouraged this growth discouraged my presence by their constant cawing. Nevermore. That was the last time I looked at a lichen with anything other than fleeting interest. The problem is that they grow so damn slowly, the crusty ones forming the circular mycelial signature on rocks, slate roofing, and tombstones at rates of a few millimeters or at most a centimeter or so per year. In the same way that mycologists eschew the charms of baker's yeast, few of us have much time for lichens. This dismissal has led to significant weaknesses in the field of mycology because close to half of all of the fungi that have been identified are ascomycetes, and more than forty percent of these—13,500 species—are lichens. The fungal nature of the lichen is most evident when it develops cups on its surface and blasts ascospores into the air. A small selection of basidiomycetes also collaborate with algae in the formation of lichens and bear tiny mushrooms rather than cups in preparation for dispersal. When the fungal partner discharges its spores, it leaves the alga or bacterium behind. This means that the fungus must capture something photosynthetic and amenable to cooperation—one of only 100 or so algae and bacteria that are found in lichens—each time it founds a colony.

Stalked like mushrooms, the head of the morel is an enormous cup covered with asci that line the gouges in its surface. The asci shoot their spores in a roughly horizontal direction, continually creating a spore cloud a few millimeters away from the surface of the fruiting body, which is dissipated by wind gusts. Speaking of delicious fungi, truffles are relatives of apothecial fungi. Since their asci develop within closed fruiting bodies that remain submerged beneath the soil, their affinity with cup fungi is not immediately obvious. Nevertheless, we can piece together a logical story of truffle evolution by making a series of sensible guesses about the alterations necessary to convert a cup into an aromatic nugget.

If we could look at clips from a movie showing the evolutionary history of truffles, we might observe the following episodes (Figure 4.5). We begin 300 million years ago in a Carboniferous forest of gigantic horsetails and pigeon-sized dragonflies. Although the plants and animals are unfamiliar, disc-shaped apothecia scattered over the forest floor look almost identical to ascomycete species we find in the woods today. Fast-forwarding 70 million years, the next movie clip shows a Permian forest of cone-bearing trees that dwarf the once massive horsetails, whose stems are now as slender as drinking straws. On the ground is a fruiting body that doesn't open out flat, but remains half buried in the soil with a ragged opening at the top. This odd apothecium developed following a mutation in a cup fungus that crippled its expansion mechanism. The mutant's asci still operate as cannons, but only those spores discharged directly beneath the opening are jettisoned into the air. Most of the ascospores accumulate inside the mutant fruiting body. Attracted by the sugary globs of discharged ascus sap, beetles crawl in through the mouth and emerge covered with sticky ascospores. One advantage that this fruiting body with the rumpled design has over the

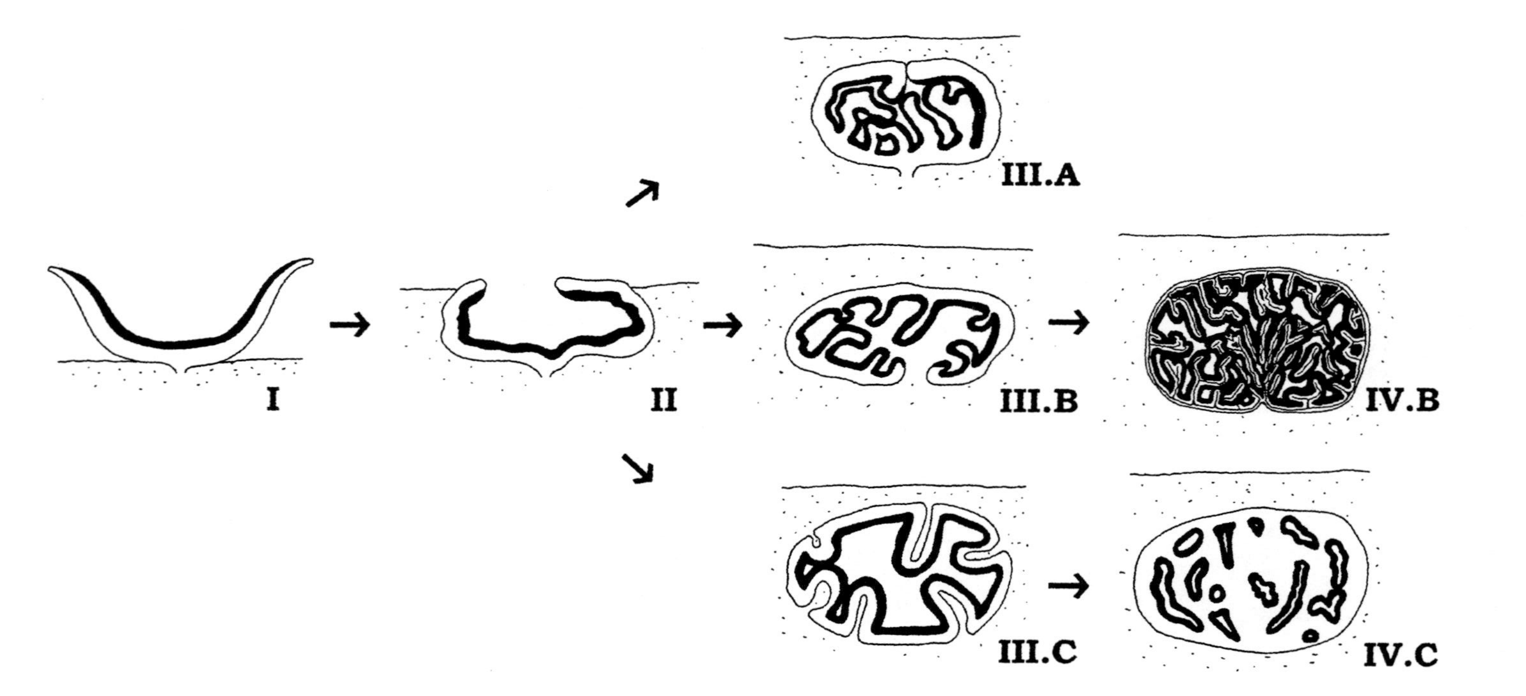

Fig. 4.5 Various hypothetical lines of fruiting body evolution in the ancestry of truffles. Each fruiting body is illustrated in section to show the distribution of the fertile tissues (indicated in black). The transition involving the formation of an inverted cup (III.B) may have occurred during the evolution of the commercial species of *Tuber* (the white truffle or black truffle shown in IV.B). Each of the diagramed fruiting bodies is exemplified by one or more living fungal species.

flattened apothecium is that it can survive for much longer in dry air than the exposed cups that shrivel after a day or so.[15]

Moving to the third clip, the soil surface in a grove of Mesozoic cycads is unbroken by fruiting bodies; then a pig-sized dinosaur with a spade-shaped nose crashes between the fronds and begins snuffling around in the soil. Unearthing fruiting bodies that look like hazelnuts, this dino-hog swallows them whole. The truffle has evolved. No longer formed in a single layer, the truffle's asci are densely packed around a network of veins that run through the interior of the fruiting body. Each ascus is a rounded bag rather than a cylinder, and contains beautiful ascospores ornamented with spines or surface ridges. Elongating an ascus, making a lid at its tip, and generating pressure for explosive discharge of spores is a waste of energy for a subterranean organism, and those individuals that suppressed these stages in ascus development flourished.[16]

The next movie clip shows a rodent nibbling a buried fruiting body. The ancestors of today's truffles survived the asteroid strike off the coast of Mexico at the end of the Cretaceous, but they did lose their spore-dispersing dinosaurs. The evolutionary modifications since then have been biochemical in nature, with the innovation of scent-producing pathways in the truffle whose products mimic the sexual attractants of mammals. Fungus and mammal have affected one another's evolutionary histories: the truffles have become more attractive to rodents and the rodents have become better at finding truffles. Interestingly, insects flit above the buried fruiting bodies—and continue to do so today—homing upon the same chemicals that have captivated them since they crawled in and out through the maw of the open fruiting body. Now that the ascocarp is buried, their role in spore dispersal, if they have any at all, is a complete mystery.

Truffle ancestors have also maintained intimate connections with the roots of various plant species during the movie. Through their mycorrhizal relationships with oak, poplar, and other trees, modern truffles obtain sugars from the plant in exchange for inorganic nutrients that they scavenge from the soil. In France and northern Italy, truffle development is encouraged through a long-term investment strategy of planting tree seedlings in fenced plots and seeding them with truffle spores or plugs of mycelium. Seven to fifteen years pass before any *tartufo* appear, which brings us to the final clip. After 300 million years of evolutionary modification, we arrive in the town square of Cortona in Tuscany and I'm sitting at a table covered with a linen tablecloth. A waiter shaves a white truffle over my plate of linguini. The translucent flakes uncurl in the warmth of the pasta. Suffused with pleasure I sip a glass of Montepulciano wine.

CHAPTER 5

The Odd Couple

There is not another genius that has come to the front in mycology in the last few years as prominently as the subject of this sketch. . . . While an old bachelor he is never as happy as rigged out in full dress and in the presence of the ladies. There is, therefore, still hope.

—Curtis G. Lloyd, *Mycological Notes* (1924)

There was a young lady named Bright,
Whose speed was far faster than light;
She set out one day
In a relative way
And returned on the previous night.

—A. H. R. Buller, *Punch* (1923)

Arthur Henry Reginald Buller stepped from a westbound train at the Canadian Pacific Railway Station in Winnipeg, Manitoba, in 1904. The thirty-year-old doctor of philosophy from Birmingham was an attractive man who wore a fashionably drooping moustache that concealed his pouty lips (Figure 5.1). Leaving his trunk in the stationmaster's office, he set off toward the downtown area. Fifteen minutes later he paused outside the McLaren, a hotel recommended at the station. Buller would live in a two-room suite at the McLaren for the next forty years, and during that time become the greatest experimental mycologist in history. His genius set him apart from all other mycologists, and most other scientists. Additional characteristics placed him among the century's foremost eccentrics.

Fig. 5.1 A. H. R. Buller, thirty years old. Photograph supplied by the Buller Library, Cereal Research Centre of Agriculture and Agri-Food Canada, Winnipeg, Manitoba.

Buller had been an undergraduate in London and obtained his Ph.D. in 1899 from the University of Leipzig, which was one of the preeminent centers for botanical investigation. He then conducted research in Munich and Naples, and taught at the University of Birmingham before migrating to Winnipeg. Buller was the first professor of botany at the University of Manitoba, a position for which he was ideally qualified. Doctoral training with William Pfeffer certified him as a modern botanist, and unencumbered by wife and children he was free to leave England. Winnipeg was undergoing rapid growth, but the prospect of an academic position in the only university in a windswept province nearly three times larger than Britain, was a strange choice for such an ambitious young scientist.[1] One advantage of the job was the autonomy offered by a young institution lacking the usual herd of disaffected faculty veterans. By contrast, Buller knew that an inexperienced professor climbing the ranks of the British academic establishment would encounter nothing but friction unless his passage was greased with an Oxbridge degree or two.

In thinking about Buller's exodus, I draw on my own experience. A few days prior to my departure from England for a postdoctoral fellowship at a prestigious American university, my new boss told me on the phone that it didn't really matter what research project I adopted. He said that my bold decision to join his lab presaged a future in science that was "virtually assured." I pictured the Swedish medallion hanging against my crisp white shirt. A more appropriate image would have been my Ph.D. diploma swirling in a toilet bowl. Wearing an imbecilic grin, I hopped on the next Virgin Atlantic flight.

Buller was a superb teacher. The young professor treated every lecture as a word-perfect performance, used his strong sense of humor to leaven the details of plant life, and displayed a genuine interest in the audience. The students nicknamed him "Uncle Regie," and many of them adored him. During a lecture on population genetics, Buller remarked that while he was a student in Germany, the rate of illegitimate births had increased by 10 percent. The real amusement here came from the students' familiarity with their professor's celibacy. Botany was a critical part of a college education at the turn of the twentieth century, especially in a province consumed with the promise of agricultural development. Once a discipline other scientists associated with flower pressing, the application of biochemical methods, cutting-edge microscopic techniques, and rigorous experimentation was moving the whole field in a respectable direction. The second half of the nineteenth century had seen groundbreaking botanical research in Germany, and Buller's appointment coincided with the scientific revolution provoked by the rediscovery of Mendel's laws of inheritance in 1900. His enthusiasm was not limited to the classroom. On Saturday mornings, Regie would walk to the farmers' market, stand on a wooden crate, and lecture to the crowd about cereal crops and the fungal diseases that afflict them. In England he would have been ignored. In Winnipeg he was a freak and became a star.

Like most professors, Buller developed a schizophrenic relationship with teaching. He taught hundreds of students each year and relished his patriarchal role in the classroom, but a deeper passion was reserved for his research on fungi. After breakfast at a downtown restaurant, the professor would hop on a streetcar at 6:00 A.M. and work in his laboratory for a couple of hours before classes began. For his first experiments in Winnipeg, he collected mushrooms from a woodpile and began investigating spore discharge. Fungal spores had first been described in the sixteenth century, but the way in which mushrooms operated as devices for their production and dispersal was quite mysterious. This was virgin scientific territory.

Buller estimated the number of spores released from different species by collecting spore deposits beneath their caps, and found that the normally invisible rain of spores could be visualized using a strong beam of light. Enclosed in a beaker, a mushroom would shower the interior with a fine dust that circulated slowly beneath the cap. Buller recognized that the behavior of spores might offer a critical test of Stokes' law as it applied to microscopic objects. This well-known law of physics relates the speed of a falling object to the viscosity of its surroundings. It was developed by George Stokes in the 1840s while studying pendulums, and helped explain the formation of clouds. To substitute spores for water droplets in Stokes' equation, Buller measured the size and density of mushroom basidiospores, and then designed a method for determining the speed at which they fell. The details of these experiments say much about Buller's ingenuity. He enclosed a thin mushroom slice in a glass-sided chamber, positioned it in a stand, and viewed the space under the gills through a microscope that was tilted horizontally (Figure 5.2). Through the microscope eyepiece, the falling spores appeared jet-black against the bright circular field of view. The microscope provided a twenty-fivefold magnification of the spores, and by calibrating the field of view with two silk threads strung across the eyepiece, the actual distance that they were moving could be measured. Finally, an electric tapper was connected to an ink pen that inscribed paper on a recording drum that spun at constant speed. The tapper was pressed as a spore fell

Fig. 5.2 Buller studying spore velocity. From A. H. R. Buller, *Researches on Fungi*, vol. 1 (London: Longmans, Green, 1909).

past the first thread, and again when it cleared the second, so that the motion of every spore was registered by two marks on the drum recording.[2] The velocity of the spores was calculated by dividing the distance between the threads by the time intervals between the pairs of markings.

In the seclusion of his laboratory Buller recorded the motion of hundreds of spores, and found that they fell somewhat faster than Stokes' law predicted. He could not explain the discrepancy. Nevertheless, the imaginative experiments were published in *Nature*.[3] During these early investigations Buller also studied the effect of humidity on spore sedimentation, the trajectory of the spores when they were propelled from gills, and their electrostatic charge. This work was described in the first volume of Buller's *Researches on Fungi*, published in 1909.[4] Six additional volumes would follow. To avoid the slashes of an editor's pen, Buller covered the publication costs for the early volumes himself.

The next three volumes of his *Researches* describe thirteen years of scientific inquiry that answered many fundamental questions about spore production and dispersal. Buller performed most of the experiments in Winnipeg, but he also worked in laboratories at Kew and in Birmingham during summer pilgrimages to England when his classes ended. In 1910, Buller observed the formation of the fluid drop—Buller's drop—at the base of the mushroom basidiospore a few seconds before discharge (Chapter 1). Once he found that the drop was carried with the spore, the problem encountered in testing Stokes' law was resolved. Fluid clinging to the surface of the spore increased its diameter by a few millionths of a meter, speeding its descent through the air. Mushroom spores indeed behaved precisely as Stokes suggested. But by then, the law had been validated by other researchers studying aerosols of mercury and wax particles.

The timing of drop expansion, coupled with Buller's discovery that the fluid and the spore traveled together, suggested that the drop was a key player in the discharge process.

> It may be that the force of surface tension is used in some way to effect spore-discharge; but exactly how I cannot satisfactorily explain. Possibly we have an entirely new principle involved in the mechanism of spore-discharge.
>
> —A. H. R. Buller, *Researches on Fungi,* vol. 2, 26 (1922)

The function of the drop in the catapult mechanism would defy explanation for more than seventy years, but Buller's hypothesis about surface tension and many other observations were critical to the ultimate solution.

Apart from his innate intelligence, Buller possessed a crucial physical resource that qualified him for his work on fungi: very keen eyesight. This is clear from his luminous drawings of fungal structures, and in the experiments on spore discharge that required hour after hour of painstaking microscopic observation. My high school biology teacher, Mr. Bigwood, claimed that work with microscopes always led to myopia. Biggie was warning me about the hazards of an obsession with biology, and although there may be no connection, I soon acquired my first glasses. Along with my nearsightedness, I became increasingly distracted by irritating bodies called "floaters," which bounced across my field of vision, marring my view of the microscopic world. I consulted an ophthalmologist. (This was two decades before another eye doctor would soothe my fears about pythiosis—Chapter 2—and is the last mention of ophthalmologists in this book.) He said that these floaters had plagued humankind forever, and with great interest, pulled an ancient book from the shelf behind his desk and swept dust from its leather covers. From the book he read me a description of a swarm of bees that cast a shadow across one victim's retina whenever he looked up, and then recounted a story about another man driven insane by a crow that appeared to fly through his vitreous humor. I bet Buller's eyes were as transparent as spring water. When I saw his photograph, I recognized those bulging eyes immediately. John Webster has orbs that positively spring from his face, and he may have spent even longer at the microscope than his predecessor.

Buller appreciated his eyes as a research tool and went to elaborate lengths to make the best use of them. Some wood decay fungi are bioluminescent, casting a dim light from their mycelia and fruiting bodies. The Jack-o'-lantern, *Omphalotus olearius*, is common in North America in the late summer and early fall. It surfaces above buried wood and forms extravagant clusters of yellowish mushrooms. When the fruiting bodies are fresh they emit a green glow that is especially evident from the gills. Although the light can be sufficiently bright to be visible outdoors at night, the best way to see luminous fungi is to take them into a photography darkroom and wait a few minutes until one's eyes adjust. Buller made a systematic study of fungal luminescence, and true to character, went to great lengths to enhance his observations. In 1923 he worked on a little bracket fungus, *Panellus stypticus*, which grows on tree stumps. In the *Researches* he describes rising at 4:00 A.M. at the McLaren and walking to his lab before sunrise so that his eyes were adapted to low light conditions. He then chronicles his observations, recording how long it took at different times of the day before he could perceive the glow from the fruiting bodies in his darkroom. To avoid the lights from street lamps he muffled his head in horse blinders during his commute, so that his eyes would arrive at the lab in peak condition. Buller took photographs using the light from *Panellus* mycelia and fruiting bodies, and studied the dependence of luminescence on oxygen availability and temperature. There were two different strains of the fungus. Collections of *Panellus* from North America were luminous, but a strain sent from England did not glow. When the transatlantic relations were paired on agar, their hyphae fused, producing a hybrid mycelium with luminous and non-luminous sectors.

Some luminescent fruiting bodies emit enough light for reading. It is claimed that in the trenches of the First World War, soldiers attached mushrooms to their helmets to avoid nighttime collisions without drawing the attention of a sniper. The function of luminescence for fungi is an enigma. For fireflies and fish, light emission serves to attract mates and lure prey, and flashes of light may allow some planktonic organisms to outwit their predators. Single-celled dinoflagellates flash in response to disturbance by fish swimming through their blooms. Crustaceans feed on these dinoflagellate blooms, and experiments show that by illuminating their immediate surroundings, they expose these predators for the fish. Fungal luminescence consumes very little energy and may have no ecological value, but some authors have speculated that glowing fruiting bodies might attract insects that would serve as vectors for spore dispersal. Their greenish light is certainly visible to insects, and the possibility that insects could augment the chief mechanism of spore dispersal by wind is intriguing. Any mycologist who has spent some time studying fungi outside the laboratory has noticed the activity of insects around fruiting bodies. Close to my home, a white bracket fungus is beginning to dissolve a fallen beech tree. Every summer I am mesmerized by the commotion of midges that flit around the stacks of spore-producing shelves jutting from the decaying trunk. Are they carrying spores? I'll have to send a student out to investigate this.[5]

Although Regie loved the company of women, he was a lifelong bachelor. His dedication to fungal research was so intense that the idea of family responsibilities may have been terrifying, although he could have found a submissive wife and maintained his erratic schedule. It is possible that he was gay, although there is no evidence of this in his voluminous correspondence, nor in the rich folklore about Buller that survives among the faculty in Winnipeg. He seems to have decided against marriage as a young man, perhaps at the time he struggled with religion. His scrapbooks of newspaper clippings, photographs, and pamphlets are preserved in the Buller Library on the University of Manitoba campus, along with his priceless collection of botanical books. He collected pictures of chapels and churches, along with photographs of clergymen, and newspaper articles and poems with a religious theme. One such poem, written by John Vanes in 1837 and entitled "Sacra Solitudo," concerns the Christian warrior who loves to be alone, and claims that "Satan, the world, and sin, are all subdued / When wrestling with his God in solitude." This must have stuck a chord with the young bachelor. Buller tried to reconcile contrary views about the origin of life, but like most biologists, was deflowered by his deepening appreciation of Darwin's dangerous idea.[6] Later, the religious theme of the scrapbooks is replaced by articles about the Great War, politics, and even eugenics, although Buller continued to attend church occasionally to indulge in his love of hymn singing.

Much of Buller's personal correspondence also survives in the Buller Library, and shows that his rejection of Christianity did not shake his vow of celibacy. Marjorie Swindell, whom he met during one of his Atlantic crossings, was much taken with Reginald. In one letter she wrote, "I do wish I could come to England with you. Maybe next year!" and in 1922, "I have thought about you often and wish I might have traveled with you again." But the 42-year-old professor did not take the bait. Buller reveled in his voyages, dropping the name of the liners, including the Grampion and Megantic, on which he had booked his next passage at every opportunity. His family was fairly wealthy and this fortune allowed him to travel to England every year. He crossed the Atlantic sixty-five times, the Second War thwarting his plan to complete an even number of trips (as I'll explain later). The pace of the journey was perfect for writing. Days of enforced contemplation on trains and the ocean liner would have helped the professor unwind from teaching and direct his attention to his *Researches*. With no opportunity for experiments, he sat in his cabin deciphering data and transcribing his laboratory notes into straightforward prose. My mentor, Frank Harold, had a quote posted above his desk in Colorado that read "For God's sake stop doing experiments and think!" Buller had the rare opportunity to pursue this sage advice.

Although devoted to mycology, the professor had many other interests. He played the piano, memorized chunks of Shakespeare and Milton, and wrote poems, some particularly awful ones concerning fungi. His relativity limerick published by *Punch* in 1923 appears in *The Oxford Dictionary of Quotations*. Buller was a polished billiards player and enjoyed sharking young men in Winnipeg's bars. He housed his own table at the McLaren. Today, the McLaren is a menacing establishment, close to the red light district. I visited Winnipeg in 2000 hoping to photograph Buller's refuge, and learned that someone had been stabbed in the hotel's basement bar the day before my arrival. Finding the

McLaren's front entrance, I peered inside. A bare lightbulb cast a yellow pall over a group of thin men swaying as they passed around a bottle of peach brandy. This didn't look promising. I considered a quick dash past the revelers and up the stairs, but reflected on my ignorance concerning Regie's room number. A cabdriver waiting outside called me over, added some details to the stabbing incident, and suggested that my quest was suicidal. As we talked, a man stumbled out of the hotel displaying a bleeding nose and an elaborately punctured forearm. Demoralized, I walked away with visions of bloodstained knives and virus-laden needles.

In the 1920s, Buller's monthly bill at the McLaren was more than 100 Canadian dollars, well beyond the means of most professors in Winnipeg. But his lifestyle was never luxurious. Living in the hotel allowed him to minimize domestic distractions, which was a lifelong imperative. By the 1930s the McLaren was losing any former luster and was described as seedy by his colleagues, but he refused to acknowledge its shortcomings. In the interests of science he was also careful to limit the time he invested in his appearance. On mushroom forays, Regie donned a raccoon coat whose thin fur was moth-eaten to the leather, and his usual scholarly attire was a four-button black jacket and striped pants cut by a tailor in Birmingham. Although fashionable when he arrived in Winnipeg, this style of suit became a most peculiar costume over the next forty years. (I'm confident that I would disturb my students by dressing like Jimi Hendrix.) A photograph of the participants at the first meeting of the Mycological Society of America in 1932 shows Buller sitting in the front row. His pant legs are hitched up, revealing one white sock, the other dark. Whether deliberate or careless, the odd socks are a modest reflection of his eccentricity.

In February 1920, Buller spent a day with Curtis Gates Lloyd in Cincinnati. Once the city's most eligible bachelor, Mr. Lloyd had ripened into a 60-year-old millionaire obsessed with fungi. He had trained as a pharmacist and held a one-third interest in the family business, Lloyd Brothers, Pharmacists Inc., headed by his elder brother John Uri. Curtis served as the field representative for the company, securing foreign sources for botanical products and searching for new medicinal plants. But at the time of Buller's visit, he had abandoned his work for the company in favor of his consuming interest in mycology. Curtis became a self-taught expert in fungal identification, and like Buller, had undertaken the publication of his own mycological work. Seven volumes of Lloyd's journal, *Mycological Notes*, appeared between 1898 and 1925. Concerned with descriptions of fungi rather than experimental studies, they also served as a bully pulpit for Curtis. He took issue with the practice of appending one's name as a "taxonomic authority" following the Latin binomial of a new species.[7] Curtis denounced a number of individuals for publishing incompetent, or even false, descriptions of new fungal species simply to inflate their own scientific reputations. To ridicule the offenders, Curtis described a number of fallacious species under the name N. J. McGinty[8] in his *Mycological Notes*. Some scientists missed the joke and requested more information on these bizarre organisms. The comedy is heightened by Lloyd's interest in gasteromycete fungi, particularly the phallic species.

Unlike a university scientist, Curtis could insult academics without fear of professional retribution. Most upset among his victims was Cornell mycologist George Atkinson. The disagreement with Atkinson stemmed from the professor's published description of a rare phallic mushroom collected in Texas. Atkinson's drawing shows a stout, hairy-tipped shaft decorated with a loose net that droops from its head (Figure 5.3 a). Atkinson erected a new genus, *Dictybole*, to encompass his Freudian monster. But, according to Lloyd, the drawing was based on a single decomposing specimen of a well-known mushroom that had been pickled in alcohol. This species, called *Simblum periphragmoides*, had been described seventy years earlier by the Reverend Miles Berkeley. It is a phallic mushroom topped with a cage that advertises the spore slime (Figure 5.3 b). When this fungus begins to disintegrate, the cage collapses and can hang from the tip: *Dictybole* is *Simblum* in death. An eagerness for publications had led Atkinson perilously close to an act of scientific fraud, although foolishness is a fairer accusation. Unfortunately for the Cornell professor, Lloyd uncovered his error and had the power and personality to impale him in print. Atkinson's shame was broadcast to everyone in the field through Lloyd's *Mycological Notes*, and Curtis kept the issue alive for the remainder of Atkinson's life. In 1910 he reproduced a fanciful illustration from an old herbal showing a

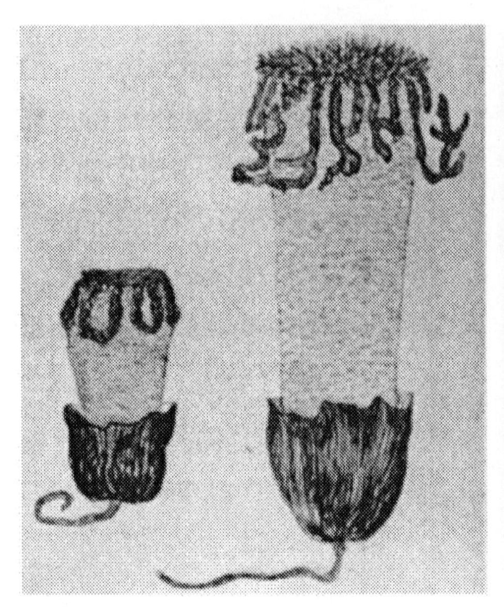

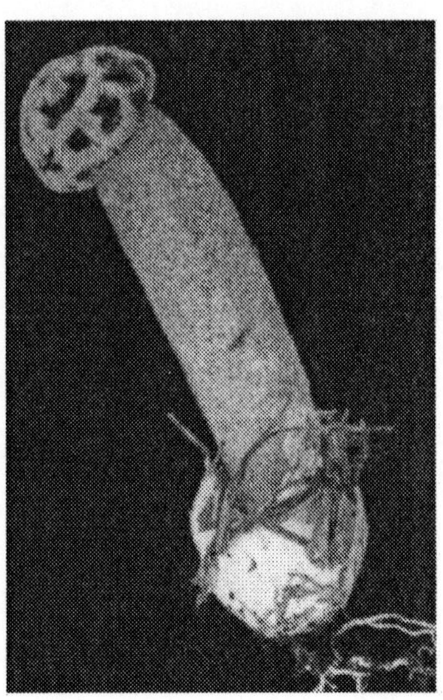

Fig. 5.3 (a) Atkinson's illustration of *Dictybole texensis*. From the *Botanical Gazette* 34, 36–43 (1902), with permission. (b) Photograph of *Simblum periphragmoides* (now named *Lysurus periphragmoides*) from S. Metzler, V. Metzler, & O. K. Miller, *Texas Mushrooms. A Field Guide* (Austin: University of Texas Press, 1992), with permission.

phallic-looking mushroom in an apparent act of ejaculation. Writing as McGinty, he suggested that this must have been a relative of Atkinson's *Dictybole*, and named it *Ætnensis* (after the volcano) *clavariaeformis* McGinty. McGinty resurfaced in 1917 to propose another imaginary genus called *Atkinsonia*.

Atkinson didn't respond to these assaults until Curtis requested a photograph of the professor. Quite understandably, he feared that Lloyd would humiliate him further in one of his biographical articles.

> I could well have pardoned the first several attacks, but the last reference to me in your notes I consider a gross injustice amounting to an insult when you propose a genus name based on my name [*Atkinsonia*] and assign it to Mr. McGinty. If you had any sense of justice or self respect, you would certainly regret that act.
>
> —George Atkinson, *Letter to Curtis Lloyd* (May 10, 1917)

Although Lloyd claimed that he meant no harm and had "nothing but the most friendly feelings" toward Atkinson, he would not apologize for McGinty. Refusing to send his portrait, the professor wrote a final defiant letter to Lloyd, signed it in a barely legible hand, and, shortly thereafter, went to his grave. Following an unsuccessful request for a photograph from the professor's widow, Lloyd illustrated the biographical sketch, now an obituary, with a portrait supplied by a colleague of Atkinson's at Cornell. Although Atkinson's pioneering photographic studies of fungi were noted, the piece lacks the usual warmth of Lloyd's vignettes.

The poor judgment of Atkinson and others did considerable damage to the field of mycology, and a hint of condescension lingers among many fungal biologists when they describe someone as a taxonomist. But Curtis Lloyd's crusade against the method of authorizing new species with the name of the discoverer was unsound. He allowed his resentment of academic pomposity and bad science to overwhelm reason. In an article entitled "The Myths of Mycology" published in 1917, in which Lloyd savaged Atkinson, he also turned his ire upon Lucien Underwood, a professor of botany at Columbia University. Underwood's transgression was to have disagreed with Lloyd's approach to taxonomy, complaining that the absence of authorities made it difficult for an investigator to trace the original descriptions of species.

Imagine that you find an unfamiliar kind of morel in the woods. Could this be a new species, something that no other person has described? This question can only be resolved by referring to the existing descriptions of morels, usually located through the author's name, and sometimes by studying herbarium specimens of the fruiting bodies on which these original accounts were based. Without this detective work, there would be nothing to stop the repeated naming of a single species. Lloyd's strategy of omitting authorities would actually inflate, rather than reduce, the catalog of erroneous new species.

> Perhaps your method is all right for you but it makes working botanists a deuced lot of trouble in following you. I agree with you heartily that the names cumber [the] literature but your system will only be available as a time saver when we have an authentic

> list of the plants of North America or of the world which can be considered a final umpire for plant names.
>
> —Lucien Underwood, *Letter to Curtis Lloyd* (May 18, 1899)

Underwood stung the amateur with the remark about "working botanists" and put forward a forceful argument. In a world in which a complete list of all species was available, there would be no need for authorities. But this is a fantasy for a mycologist. At least a million fungi await discovery, and many of the 74,000 or so that we have identified break most of the rules that are supposed to define a species. A single basidiomycete species can encompass an unsettling amount of genetic diversity among its individuals and produce fruiting bodies of varied shapes and colors. When we add the specter of hybrids between species, and the formation of multiple spore-producing forms in a life cycle, it becomes clear that we'll never have a definitive global catalog of fungi. But although Lloyd never answered Underwood's critique, he remained convinced that his own argument was solid. He could say what he liked in his *Mycological Notes*.

For many years, Curtis lived at 309 West Court Street on the ground floor of the Lloyd Library. The library was established by the Lloyd brothers as an archive for botanical and mycological works, and also as a research laboratory and museum. The private collection remains in downtown Cincinnati, now housed in a brick building on the edge of one of the poorer neighborhoods in the city. Such is its detachment from its surroundings and the value of its contents, that this bombproofed island reminds me of the church in Aksum, Ethiopia, said to contain the Ark of the Covenant.[9] The half of this book that was not written in my shed, was composed in the Lloyd.

As a young man, Curtis enjoyed a lively social life, and does not seem to have suffered with images of solitary Christian warriors (Figure 5.4). Along with the Lloyd Library's collection of his mycological correspondence are files stuffed with notes from young women. These range from cards with one or two provocative sentences, usually thanking Curtis for a delightful evening, to photographs and poems from more enthusiastic acquaintances. He was a handsome man and very wealthy, and had lots

Fig. 5.4 Curtis Lloyd relaxing in hammock with unidentified admirer. Photograph courtesy of the Lloyd Library, Cincinnati, Ohio.

of free time. He would have qualified as the great catch of the 1890s in Cincinnati, but like Buller he never married.

Nellie Julia Thompson worked at the Ohio Phonograph Company and spent many evenings with Curtis dancing, playing cards, and attending the theater and opera. Suffering the illusion that she might become Mrs. Lloyd, Nellie endured a tortuous three-year flirtation with Curtis. She wrote to him after every date. Early on, she quipped with Curtis about his bachelorhood, and teased him with the knowledge that other men were courting her. Curtis responded with gifts and invitations, but then retreated. Nellie addressed Curtis as her "Dear Brother" and "Bachelor Friend," and never admitted an interest in marriage, but her letters take on the hint of a pursuer. She bought ties for Curtis, referring to herself as his "purchasing agent," but noting that "you never do like anything I get or anything that I do. For you always turn every thing to ridicule" (Nellie Thompson, letter to Curtis Lloyd, July 23, 1891). Later, in the same letter she wrote, "Mr. Lloyd, I have come to the conclusion that you are Rudeness Personified, for you always desire other companionship than that of the lady with you, but you are pardoned somewhat on account of the fact that you are a crusty old bachelor." Curtis agreed that he was a crusty old bachelor but disputed the accusation of rudeness. He had a habit of drawing lines in the sand, setting dates that would determine his withdrawal from the society of women in favor of the refuge offered by his club. Nellie alluded to one of these resolutions as his plan to retire from "The Charms of Feminine Loveliness," although she was mistaken in thinking that Curtis intended a general retreat. She was one of many female friends. In 1894, Curtis received a letter from a Cincinnati socialite, mistakenly addressed to Mrs. C. G. Lloyd, and as a twisted joke sent it to Nellie. In the final letter from Nellie that Curtis kept, she wished him a happy birthday and appealed to him to forego his club for one evening and call on her. He never deserved her.

Originally, I went to the Lloyd Library to research Buller's writings on spore discharge for an article commissioned by the journal *Mycologia*. I continued to visit the library for several years, often thinking about Buller without knowing that he had been a friend of Curtis and that their letters were shelved a few feet from my desk. Bound by their mutual authority in the mycological community, common obsession with fungi, and similar lifestyles, their brief meeting in 1920 was a rare delight for the bachelors. Lloyd's frankness and lack of pomposity would have been a breath of fresh air for Buller, and the professor's interest in experiments rather than species descriptions allowed them to disengage from Lloyd's crusade against taxonomists. After their meeting, Lloyd requested a photograph of Buller for his *Mycological Notes*, and they corresponded about various fungi until Lloyd's death. In almost every letter, the friends refer to that winter's day in Cincinnati.

Curtis had suffered from diabetes for many years, and with little in the contemporary pharmacopeia to prolong his life, he died of complications from the disease in 1926. In his obituary, *The Cincinnati Enquirer* wrote: "Funeral services will be dispensed with, his body cremated and the ashes dissipated all without ceremony. A letter of instructions, sent to his executor and referred to in his will, stipulated that his remains be forwarded 'without discourse, ceremony flowers or funeral services of any kind to the nearest crematory and the ashes dissipated. No one is to be present except crematory officials.'" His ashes were scattered over one of the gardens in the family estate in Crittendon, south of the Ohio River in Kentucky. Curtis had designed his own tombstone, and erected the slab of rough-hewn granite five years before his death. It bears the following inscription:

> Curtis G. Lloyd. Born in 1859. Died 60 or more years after. The exact number of years, months and days he lived nobody knows and nobody cares. Monument erected by himself, for himself, during his own life, to gratify his own vanity. What fools these mortals be!

Aside from its general stab at narcissism, Lloyd's epitaph is an enduring attack on the egotism of scientists who fight to attach their names to anything that might afford them the briefest of spells in the dim limelight accorded by the profession. So very few biologists gain name recognition beyond their own institutions or narrow fields, and the rare superstar is almost always envied and despised. Lloyd could not forgive the peacocks and left his final rant in stone.

Buller's researches continued and his reputation in the scientific community grew. His talent as a lecturer led to frequent invitations from universities in North America and England. He delivered six evening lectures at Northwestern University in 1927 under the banner, "Recent advances in our knowledge of the fungi, or the romance of the fungi life. Illustrated by lantern slides, models and specimens." In addition to his continuing fascination with spore release mechanisms, Buller was making important contributions to the study of fungal genetics. He investigated the sexuality of ink-caps and the processes of hyphal fusion and nuclear migration that precede mushroom formation. Contemporary molecular research on mating types and the genetics of compatibility has built on his experiments. But even more significant was his work on the rust fungi that parasitize cereal crops. As I'll explain in the final chapter, rusts have evolved astonishingly complex life cycles involving as many as four types of spore and the infection of two different plant species. Although much of the life cycle had been cracked by earlier researchers, the sexual process was not unraveled until the 1920s. The missing information was significant, because sex between compatible rust mycelia transmits the virulence genes that allow these fungi to attack new wheat varieties. Buller had a hunch that flies could be responsible for transferring sex cells from one rust strain to another, in the same way that insects act as pollinators of flowering plants. His idea was correct, and by advising the plant pathologist John Craigie to study insects as they visited infected leaves, Buller played a leading role in solving the puzzle of rust sexuality.

He was elected to the Royal Society of Canada and later to the Royal Society of London, adding to the ever-growing appendage to his name which he cited as "A. H. Reginald Buller, F.R.S.; B.Sc. (Lond.); D.Sc. (Birm.); Ph.D. (Leip.); LL.D. (Man., Sask., and Calcutta); D.Sc. (Penn.); F.R.S.C., etc.; Professor of Botany at the University of Manitoba." Buller had become a whale in a little pond, and all across the campus the disaffected were sharpening their harpoons. An earlier collaboration with the physicist Frank Allen led to accusations of stolen ideas, and other colleagues boiled with resentment over old arguments. The campus became poisoned by stories about the old mycologist's transgressions. Even now, with all of the enemies turned to dust, some of the Winnipeg biologists bristle at the mention of his name. Buller retired in 1936, assuming that he would have continued use of his office and laboratory. Unfortunately, the university comptroller turned Buller out of his office, confining him to a single desk in a crowded laboratory. A few days later, his leather furniture was dumped outside the building and became soaked in a rainstorm. The cruelty of some of the administrators had thirty-year-old roots in their fight with Buller over plans for a new site for the university campus. In a final snub, the faculty club refused to accept the gift of his treasured billiard table. In 1939, Buller moved back to England, intending to spend his retirement in Birmingham, with frequent visits to Kew. But later that year after the outbreak of war, he became stranded in New York City while attending the International Microbiological Congress. Forced into exile, he returned to his adopted home in Winnipeg and continued writing the *Researches*.

In January 1944, after suffering from severe headaches for several months, Regie checked himself into the Winnipeg General Hospital.

When the seriousness of his condition became apparent, he directed his concern to his unpublished research. He bequeathed his manuscripts to Kew and his extraordinary botanical library to the government's Dominion Rust Research Laboratory on the University of Manitoba campus. Buller and his tumor died in 1944. The brain slice pictured in his autopsy—among the more surprising items in the Buller Library—shows a massive growth distending the right frontal lobe, a glioblastoma that destroyed the finest brain ever to have spent a lifetime making sense of fungi. Buller wished to be cremated, and since Winnipeg lacked a crematorium, his body was transported to Minneapolis. The ashes returned in a copper urn and were kept on a shelf in a faculty member's home until the special library was built in the Rust Laboratory. The urn was left open on social occasions, allowing Buller's ageing adversaries to add their cigarette ashes to the great man's bones. Fourteen years after his death, the urn was set behind a bronze marker in the library wall, and in 1963, when guilt had replaced venom, the science building on campus was renamed the Buller Biological Laboratories.

CHAPTER 6

Ingold's Jewels

Discovery consists of seeing what everybody has seen and thinking what nobody has thought.

—Albert von Szent-Györgyi

I spent my childhood in the Oxfordshire village of Benson. In the 1960s the community experienced a population explosion and new homes gobbled up the surrounding wheat fields and dairy farms. Forty years later, Benson has reverted to a sleepier existence, but returning as a middle-aged American academic (there's an attractive image), I can't decide whether paradise has been regained or lost. It had been such a lively place.

Benson sits on the north side of the River Thames or Isis, to whose flow it contributes a chalk stream or brook that bubbles through the village along a stone-lined course. The brook water is clear as crystal and flashes with trout and stickleback. It teems with crayfish, insect larvae, and a profusion of other invertebrates, and here and there flourish patches of the pepperiest watercress available to humanity. And invisibly, the water carries fungal spores of incomparable beauty. If someone had told me about aquatic fungi when I lived by the brook, I would not have been impressed. As a boy, I daydreamed of piloting one of the jet fighters that streaked over the village, and later, my teenage imagination was commanded by an even less likely fantasy involving an angel of incomparable beauty called Jane. But years after my infatuation with this schoolgirl dissolved, and now that the thrill of watching aircraft has been replaced by the chore of boarding them, I can't stop thinking about those spores. Time and a microscope can change one's perspective remarkably.

Water and life are inseparable, and while nothing that living things do is imaginable without this miraculous solvent, there is a particular intimacy to the way water commands fungal processes. In addition to its routine biochemical functions, fungal water acts as an internal skeleton in mycelia and mushrooms, powers invasive growth, and is indispensable for drop-driven catapults and other gadgets for launching spores. The continuous fungal demand for water is evoked by the luxuriance of fruiting bodies in humid forests and a general abundance of all things mycological in every wet location. The above-ground signs of fungi enjoy universal recognition, but mycologists are also aware of the abundance of truly aquatic species that live and reproduce under water, erasing nature's refuse in freshwater habitats and colonizing the timbers of sunken galleons. This chapter is concerned with the biology of a cross-section of the aquatic fungi, from species with star-shaped spores that are carried passively by moving water, to others that qualify as masterful swimmers.

In fast-flowing creeks, foams froth around half-submerged branches or at the bottom of waterfalls. Industrial activity may be blamed for the foulest froths, but other foams are a natural phenomenon caused by chemicals released from decaying leaves. These fatty substances have the properties of detergents and form bubbles when the water is churned violently. Cakes of foam trap and concentrate the same kinds of marvelous spores all over the world: some are star-shaped with thin limbs connected to a central hub, others are crescent-shaped or sigmoid (an elongated S twisted into an extended helix), and a few combine these features and look like the animals created by those balloon sculptors that haunt children's hospitals and carnivals (Figure 6.1). These are the conidia of Ingoldian hyphomycetes, named for their discoverer, Cecil Terence Ingold (Figure 6.2). As a young professor in Leicester in 1938, Ingold found them in foam that collected in "a little, alder-lined, babbling brook" close to his home.[1] After months of research he concluded that the spores were formed by a hitherto unknown group of aquatic fungi that were instrumental in leaf decomposition. There had been a few earlier reports of aquatic spores with long appendages, but most biologists ignored them when they appeared in water samples, or misidentified the spores as protozoans.[2] When Ingold described his

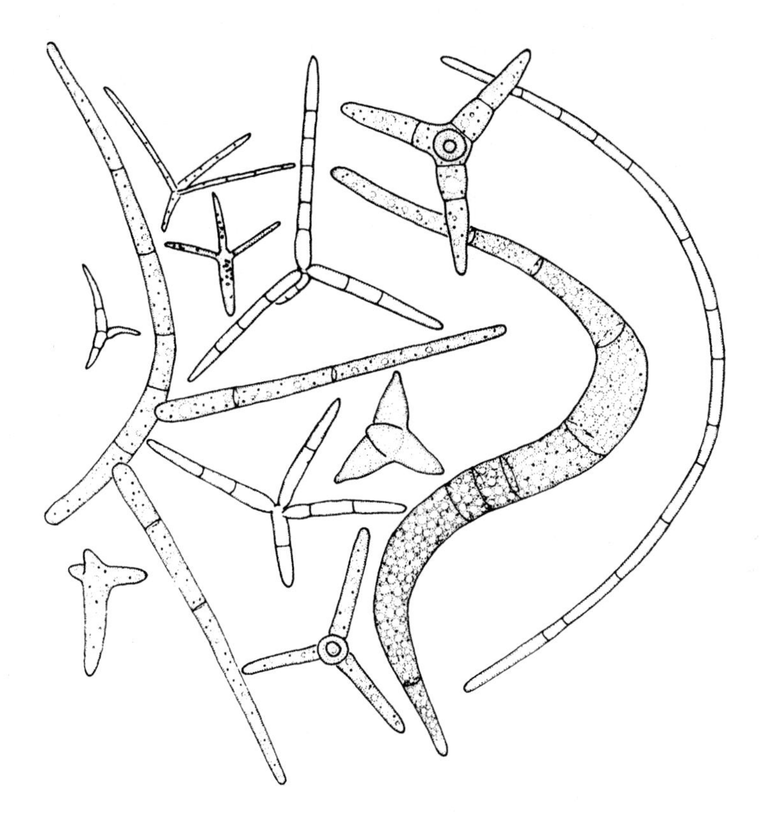

Fig. 6.1 Variety of Ingoldian spores. From J. Webster, *Introduction to Fungi* (Cambridge: Cambridge University Press, 1980), reprinted with permission.

spores at a meeting of the British Mycological Society, a twit from Cambridge University dismissed them as hairs from plant leaves. Sixty years later, more than 300 species of Ingoldian fungi have been identified and they are the subject of more than 1,000 scientific publications.

To mycologists, Ingold's discovery stands comparison to the first scientific reports of elephants, rhinoceros, and other African mammals, or descriptions of the marsupial faunas of Australia or Madagascar. A wholly new type of fungus had been found, a rich mine of species that had always, but until then invisibly, fashioned their spores in creeks or streams on every continent. Henry Descals (Webster's research associate who accompanied us on our phallic mushroom expedition at the beginning of the book) is an authority on Ingoldian fungi who has been working on a definitive description or monograph of the group for more than

Fig. 6.2 C. T. Ingold photographed in 1959.

twenty years. Whenever it seems that his catalog is nearing completion, a horde of new Ingoldians is uncovered and the date for publication is extended. To isolate the spores from the surface of decaying leaves, Henry plucks his long eyelashes and glues them to cocktail sticks. Eyelashes have strong, pointed tips, making them perfect instruments for manipulating these microscopic jewels. After working in England with Webster, Henry Descals returned to Spain to pursue Ingoldians in waters closer to his family's lemon grove. When he said goodbye, he remarked that his years in Exeter had been the happiest of his life. In part, Henry was acknowledging the satisfaction that had come from his research, something that is eliminated from the lives of most scientists by the necessity of bottom-clenched petitions for funding and the scramble toward the next rung on the ladder. Nevertheless, if you haven't had the opportunity to explore a question for its own sake, to rely on your own creativity to unravel something that nobody in the history of the universe has ever understood, then you have missed one of the finest experiences in life. I suggest you quit your job immediately and get to the nearest university as fast as you can. But tread carefully. All that clenching has created the most extravagant bastards and you must choose your research supervisor very wisely.

Ingoldian spores reach concentrations of 20,000 per liter of water in late fall and early winter, so someone splashing around in Benson Brook is exposed to millions of them. Fortunately, no human infections by Ingoldians have been reported. As the spores flow downstream, glittering in shafts of sunlight, they colonize submerged leaves and spawn mycelia that soften and macerate the plant tissues by secreting cellulose-degrading enzymes. Hyphae thoroughly insinuate themselves in the decaying leaves, raising the protein content of the debris as dead plant is transformed into living fungus. This microbial processing of the leaves is a boon for tiny crustaceans that flourish on the resulting compost. Ingoldians exit their substrate by sprouting forests of hyphae at the leaf surface and forming a canopy of spores. The development of the star-shaped or tetraradiate spores of the fungus *Actinospora* is an impressive performance. A hyphal apex swells to form the hub, from which four buds emerge and then elongate into arms. All the arms extend at the same rate until the huge spore spans a diameter of 0.5 millimeters. At this size, single spores are easily visible with a magnifying glass or hand lens. Although it is not listed in *The Guinness Book of Records*, this is the largest spore formed by any fungus. (I think this is an unforgivable omission in light of the fact that the publisher has expended ink on a Frenchman who has eaten more bicycles than any other human being. Incidentally, his name is Michel Lotito and he has swallowed eighteen.)[3] Once mature, each *Actinospora* star separates from its hyphal stalk and is drawn off into the water, arms outstretched to embrace the next leaf.

Perpetuation of the genes housed in the star-shaped spores depends upon attachment to fresh leaves that drop into the water. Some mycologists have contended that the advantage of an extended form might be its effectiveness at slowing the speed at which spores settle in water. A slow descent would be advantageous, because it would increase the opportunity for colliding with a submerged leaf before plopping into the sediment on the bottom of the creek. But meticulous experiments have demonstrated that spores with microscopic arms fall through the water column as fast as spherical spores with similar mass.[4] Water represents a viscous medium for microscopic particles—with or without arms—and slows their descent to rates of millimeters *per minute*, versus millimeters *per second* in air. If cell appendages were really effective at slowing sedimentation, spores with elaborate shapes would be anticipated in terrestrial fungi, because the faster rate of descent in air places an even greater premium upon any structure that acts as a brake and thereby favors horizontal transport. But mushroom spores and airborne conidia are always compact, supporting the conclusion that any appendage on a microscopic spore adds unwelcome mass and speeds descent. The unusual shapes of aquatic spores require an alternative explanation.

The most compelling answer is supported by a branch of mathematics called search theory, which was developed originally for antisubmarine warfare.[5] Not surprisingly, analysis shows that the probability of hitting a target increases in proportion to the length of the "search vehicle," and elongated spores seem to be very effective for organisms engaged in a passive hunt for solid materials in a liquid environment. A spore shaped like a star has a much higher probability of encountering a submerged leaf than a spore of equal mass with a condensed form. Because the appendages of tetraradiate spores are no thicker than hyphae, these decorative structures span an enormous area (for a microorganism) with a minimal investment of cytoplasm. There is a second advantage to the tetraradiate shape. When the tip of an arm strikes a solid surface, water movement forces the spore to pivot around this point of attachment[6] until an additional pair of arms lands on the substrate to complete a three-point landing. Upon contact, the end of each arm swells and secretes an adhesive that cements the tripod to the leaf. The swellings at the ends of the appendages produce infection hyphae that penetrate the leaf, so that a single spore attacks its food from the corners of a triangle. By tumbling in the water flow, Ingoldian spores with a sigmoidal shape also explore a large volume of water. When they encounter a surface, sigmoid spores often roll sideways, or somersault end over end until a single firm attachment is made. They then secrete adhesive at the second point of contact, either at the other end of the spore or at some intermediate point, to prevent further disturbance.

Spore formation is not restricted to submerged locations. Ingoldians are also found on plants that overhang creeks, and enter the water during rainstorms. Their appearance above water seems incongruous for an aquatic organism that feeds on rotting leaves, but this is explained by their mermaid-like life cycles. Mycelia that form tetraradiate and sigmoid spores are the asexual phases or anamorphs of fungi that produce sexual spores in other habitats (Chapter 4). The enormous spores of *Actinospora* are products of an ascomycete cup fungus, and other Ingoldians are the asexual manifestations of basidiomycetes and different types of ascomycete fungi. Certain zygomycetes that parasitize aquatic insects also produce tetraradiate spores in creeks. The fabrication of spores with a tetraradiate morphology by unrelated fungi suggests that these shapes evolved independently to meet similar environmental challenges. This is another mycological case of evolutionary convergence (the independent derivation of hyphae by stramenopile fungi and mushroom relatives was discussed in Chapter 3).

We have seen that Ingoldians are most abundant in the turbulent, highly oxygenated water of shallow creeks. A different group of fungi called the aero-aquatics specialize in leaf decomposition in filthy water and develop ornate spores shaped like barrels and cages that trap air bubbles. Barrel shapes are constructed by hyphae that grow in a tight helical path, and cages form when hyphae aggregate and branch repeatedly. Their air bubbles resemble globules of mercury when the spore is submerged and the resulting buoyancy lifts the spores from their underlying mycelium and allows them to drift along the surface of the water. In common with the Ingoldian fungi, the conidia of aero-aquatics are clones of the parent. They are generated without sex. But this is not true of all aquatic spores. Highly modified sexual spores are characteristic of basidiomycete and ascomycete mycelia that colonize driftwood in the sea. Fruiting bodies of *Nia* are tiny orange globes covered with hairs and filled with basidia. Because these resemble miniature puffballs, *Nia* is classified in that ragbag of strange species called the gasteromycetes. The fungus has been studied most intensively on timbers recovered from the wreck of the Tudor warship *Mary Rose* that sank off the British coast in 1545. For four centuries the ship was preserved by ocean sediment, but as her hull was raised it was attacked by *Nia* and other fungi. Decay was encouraged inadvertently during early phases of restoration when the timbers were wrapped in polyethylene sheeting, which maintained the sodden conditions favorable for fungal growth.

In accordance with its classification as a gasteromycete, *Nia* has dispensed with the surface tension catapult mechanism of spore discharge. This is a safe prediction for any aquatic fungus because Buller's drop cannot develop under water. Instead, swarms of basidiospores are shed from *Nia* when its fruiting bodies swell with mucilage and burst. The shape of the spore is familiar: each radiates four or five appendages from a swollen central hub. These saltwater basidiospores were misidentified as conidia when they were discovered in the 1950s and could be easily mistaken for Ingoldian spores. Again, evolutionary history repeats itself through convergence. Marine ascomycetes also outfit their spores with appendages and increase their "search area" with mucilage rings and tails that aid attachment when they encounter something solid. There has been no need to modify the terrestrial method of discharging ascospores because the pressure-driven cannon functions perfectly well in the ocean.[7]

Like Ingold's conidia, aquatic basidiospores and ascospores rely on water movement for locomotion, and impaction on a food source is a matter of chance whose likelihood is largely determined by the relative concentrations of spores and food. Other fungi are more aggressive and target their substrates with motile cells called zoospores. The zoospores of chytrid fungi drive themselves through water using a single tail called a whiplash flagellum that undulates from its base to its tip (Figure 6.3 a). Flagellate cells are found in most eukaryotes, with the exception of nematode worms, flowering plants, and everything in Kingdom Fungi apart from the chytrids. Human flagella (known as cilia when there are lots of them attached to a single cell) push sperm cells toward eggs, circulate mucus in our lungs, and enable us to hear. The breakdown of fuel molecules in mitochondria powers flagellar motion, while inside the ear, modified cilia remain static until disturbed by sound waves. As a teenager, my cilia were distressed at a concert by gentlemen of the performing arts known as Motorhead whose lead singer looked capable of giving Michel Lotito a run for his money. Judging by the extent of my injuries, any chytrids in the audience would have been obliterated.

Chytrids are found everywhere in freshwater habitats and wet soils. Recently, this group of fungi achieved fame by parasitizing frogs and toads. For the past decade, biologists have been concerned by declin-

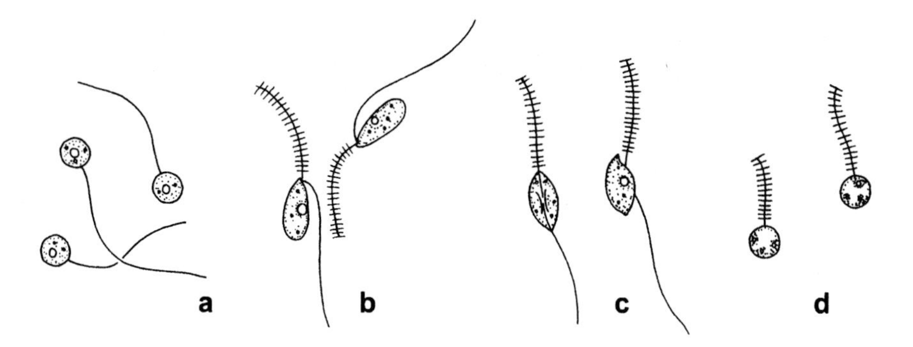

Fig. 6.3 Different types of fungal zoospore. (a) Chytrid zoospore with single smooth flagellum. (b) Primary zoospore of the oomycete *Saprolegnia*. Two flagella attached to narrow end of cell. (c) Secondary zoospore of *Saprolegnia*. Flagella emerge from cleft in the side of the spore. (d) Zoospore of hyphochytrid fungus with single flagellum clothed with filaments. The forward-directed flagellum in b, c, and d is decorated with lateral filaments. These are invisible unless an electron microscope is used to view spore.

ing numbers of amphibians in diverse locations, including places where environmental damage appears minimal. No definitive cause has been identified. Ozone depletion in the stratosphere and the resultant increase in intensity of ultraviolet radiation is a logical consideration, because amphibian eggs mature in exposed locations and are exquisitely sensitive to UV damage. According to the worst scenario, amphibians are better canaries than canaries (for human purposes), and unless the ozone shield repairs itself, we may be forced to become nocturnal. Whether ozone is a factor or not, scientists have blamed different fungi for the demise of the amphibians. First came a multinational group that discovered spore-producing capsules or sporangia buried in the skin of sickened frogs. Each sporangium was furnished with a chimney that opened at the skin surface and spat chytrid zoospores into the water. The chytrid was described in a paper in the *Proceedings of the National Academy of Sciences* in 1998 and resulted in a flurry of newspaper articles.[8] Later, the pathogen was named *Batrachochytrium dendrobatidis*. In an effort to stake their claim to solving the mystery, the authors indicted the chytrid as the cause of the amphibian casualties but did not pay much attention to the obvious possibility that the infections were a secondary consequence of other sources of stress. Four years later, another research group discovered that eggs of western toads in the Pacific Northwest were succumbing to lethal infection by an oomycete fungus called *Saprolegnia ferax*. Their work was published in a superb article in *Nature*[9] which concluded that eggs became vulnerable to colonization when the depth of their ponds decreased and they were exposed to damaging levels of ultraviolet radiation (specifically the UV-B wavelengths that initiate skin cancer in humans). Reductions in pond depth were linked to changes in rainfall patterns that have been caused by warming of the Pacific Ocean. *Saprolegnia* is acting as a classic opportunist. Based on this study, toad disappearance seems to be caused by the following chain of events: global warming, decreased rainfall at high elevations, UV-B damage (perhaps intensified by ozone thinning), and finally, fungal infection. These types of complicated interactions between climatic conditions and animal health offer a much more compelling explanation for the epidemic of amphibian disappearance than any single factor.

Besides infecting frog skins, chytrids do lots of interesting things. Some are pathogens of plants. Zoospores of *Olpidium brassicae* swim through capillaries of water between soil particles and attach themselves to root hairs of cabbages, lettuces, and other leafy vegetables. The entry mechanism is different from the usual production of an infection hypha. *Olpidium* spores puncture the root hair with a short tube, through which their cytoplasm is siphoned into the host cell. The fungus swells into an oval body or thallus and never produces hyphae. After two or three days, some of the thalli develop exit ducts and *Olpidium* reemerges as hundreds of zoospores. Often, the effects of the chytrid infection are not noticeable, but *Olpidium* causes far greater damage by acting as a courier for viruses. Viral particles attach to the surface of zoospores as they swim through contaminated soils. When the fungus infects the plant, it transmits rod-shaped, RNA-carrying viruses, including the agent that causes lettuce big-vein disease. Another chytrid, *Synchytrium endobioticum*, causes potato wart disease and is a serious agricultural problem. Horrible masses of dark-brown tissue swell from tubers colonized by the fungus, and the infected crop is inedible. Once established in soil, the disease is not easily eradicated because *Synchytrium* has thick-walled spores that can survive for more than forty years. In Newfoundland, crab shells are a waste product of local fisheries and offer promise for controlling the disease when they are crushed and applied to soil. The action of this unique soil amendment is subtle. In common with other mushroom relatives, the walls of chytrids contain microfibrils of chitin, the same compound that comprises the crunchy parts of insects and crustaceans. By adding large quantities of chitin-rich crab shell to the soil, farmers encourage an abundance of bacteria and fungi that degrade chitin. The destruction of the chytrid resting spores is a byproduct of this radical change in soil ecology.

Chytrid species also infect mosquito larvae and other aquatic insects, algae, and oomycete fungi. Anaerobic chytrids are poisoned by atmospheric oxygen. Some of them live in the largest of the four stomach chambers—the rumen—of large herbivores, including cows. In the unappetizing soup that churns inside a cow, chytrids decompose plant fiber or cellulose. By altering the chemical composition of this material, anaerobic chytrids facilitate the growth of other microorganisms critical to the digestive mechanism, including the bacteria responsible for methane production. This food processing activity parallels the ecological role of Ingoldian fungi in creeks.

Like conidia and chytrid zoospores, each zoospore produced by an oomycete water mold carries exact copies of the chromosomes present in the nuclei of the parent colony. Zoospores carry this inheritance for a few centimeters, or as far as one meter, and are then capable of establishing a new colony. This clonal way of life, punctuated with rare sexual encounters, is a strategy that has proven very effective for oomycetes. Water molds have probably been releasing clouds of zoospores for hundreds of millions of years, surviving the catastrophic asteroid impacts that obliterated the majority of terrestrial and aquatic species at the close of the Permian and Cretaceous periods.

Oomycete water molds manufacture two types of zoospore with paired flagella (Figure 6.3 b,c). The primary zoospore has a teardrop shape, with the flagella anchored in its pointed end. The secondary zoospore resembles a kidney, and its flagella wave from a groove set in its indented side. In both cell types the flagella are oriented in the same way. One is always aimed ahead of the spore and is clothed with fringes of lateral hairs called mastigonemes; the other is smooth and points rearward. The posterior flagellum oscillates and creates forward thrust like the tail of a tadpole. At first sight, the simultaneous undulation of the hairy flagellum would appear to counteract the action of the tail motor. Initially it was thought that it may have guided spore motion by acting as a rigid fin. Careful observation with good microscopes soon disproved this idea, and the effectiveness of the forward-directed flagellum is also illustrated by the fact that zoospores of a small group of fungi called the hyphochytrids swim with nothing else (Figure 6.3 d). The paradoxical orientation of the frontal flagellum is explained by the fact that its hairs reverse the direction of thrust generated when waves pass along its length. The hairs pull the spore through the water somewhat like the arms of a swimmer doing the breaststroke. The combined push from behind and pull from the front propels the oomycete zoospore at speeds of up to one meter per hour, or 27 times its body length (ten-millionths of one meter, or 10 µm) every second. This is very impressive. To compete with the zoospore, I would need to swim faster than 160 kilometers or 100 miles per hour (which I cannot accomplish on a full rumen). However, this comparison ignores the profound effect of size on the mechanical behavior of objects in water. Unlike the human swimmer who is capable of gliding for short distances (our bodies have inertia), zoospores stop dead the instant flagellar movement ceases because they are utterly constrained by the viscosity of their surroundings.

The production of zoospores involves a biochemical decision by the fungus, among whose cues starvation is king. In the nineteenth century a German botanist, Georg Klebs, studied the nutritional requirements of water molds. The salient message of three scientific papers published between 1898 and 1900 (filling 270 printed pages) was that spores developed when the fungus exhausted the sugars in its growth medium. Much of the process of sporulation in oomycetes remains mysterious, but its beauty under the microscope has captivated mycologists since the first detailed descriptions appeared early in the nineteenth century. To understand how this works, we must begin by thinking about the structure of the mycelium. The large hyphae of oomycete water molds extract food from dead insects floating in ponds and then protrude from the surface of the corpses into the water. Unless the microorganism locates another meal it will starve, and its zoospores are dispatched as search vehicles. The first sign that the organism has entered the spore-producing pathway of development is that its hyphal tips cease extending and begin to darken (Figure 6.4). The color change is caused by the transfer of cytoplasm from older parts of the colony into the tips. Once each hyphal apex is primed with cytoplasm, it is isolated from the rest of the colony by a thick cross-wall or septum. The resulting multinucleate cell is called the sporangium. During the next hour its cytoplasm is reorganized or cleaved into spores, which are then shot into the water.

Early in the developmental process the smooth tip of the sporangium is refashioned into a nozzle. This requires the secretion of wall-loosening enzymes at the extreme apex that permit extension of the wall. At the same time, the outlines of individual spores begin to appear, as nuclei and portions of their surrounding cytoplasm are chopped out by flattened membrane-bound envelopes that score through the cytoplasm. Under the microscope, the edges of these membrane sheets look like thin canals filled with clear fluid. After 45 minutes or so, the observer is treated to the first dramatic event in the sporulation process (dramatic, like erotic, is a subjective term): the canals separating the developing spores disappear, and the lumpy appearance of the sporangial contents is replaced with the facade of unpolished marble. (You have to see this for yourself; the metaphor is apt.)

This incident marks the breakage of the membrane that encircled the entire inner surface of the sporangium and was derived from the original hypha. When this happens, the spores become squashed together as the sporangium loses pressure and collapses. The marbled pose is worn for two minutes, a period of utter stillness under the microscope, before the individual spores become visible again and separate from the inner surface of the sporangial wall. In *Saprolegnia*, each spore sprouts its flagella and these begin to lick against the inner surface of the sporangial wall. At the tip of the sporangium, the inner surface of the nozzle is caressed by the closest spore. This cell then moves into the breach, the nozzle wall bulges and ruptures, and the spore is shot silently into the water. The other zoospores follow as a stream, lining up along the central axis of the sporangium, moving in continuous file into the water (Figure 6.5 a). As each spore is ejected it seems momentarily stunned, and then its flagellar engines are engaged and the spore streaks from the

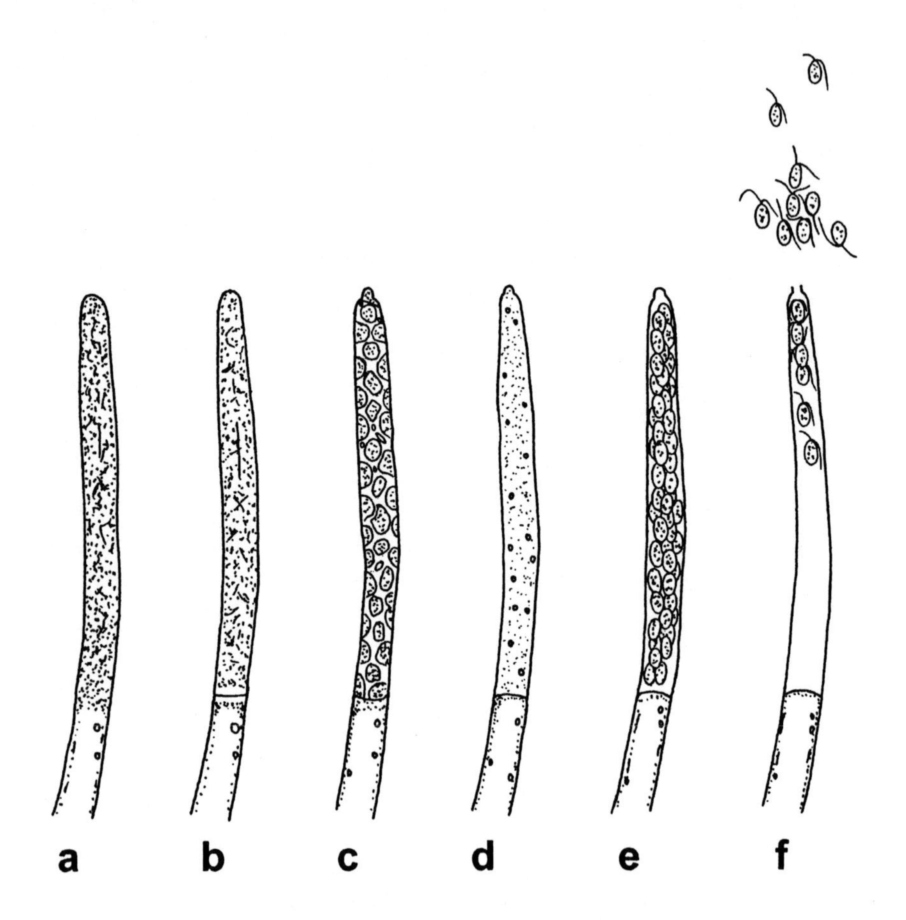

Fig. 6.4 Diagram of sporangial development in *Saprolegnia*. (a) Hyphal apex charged with cytoplasm. (b) Young sporangium isolated from mycelium by septum. (c) Beginning of cytoplasmic cleavage into spores. Nozzle has formed at tip of sporangium. (d) Homogeneous phase when sporangium loses most of its pressure. (e) Separation of individual spores. (f) Sporangial emptying.

microscope field of view. Up to 120 tearshaped primary spores are expelled within 30 seconds, leaving an empty sporangium in their wake.

I watched this process hundreds of times as a doctoral student with John Webster. Because the appearance of the sporangium changes in a predictable fashion, the moment of discharge can be estimated with great accuracy. In my laboratory classes I have disarmed the least attentive student by glancing at a specimen and saying with justified superiority that the spores will be expelled from a particular sporangium within the

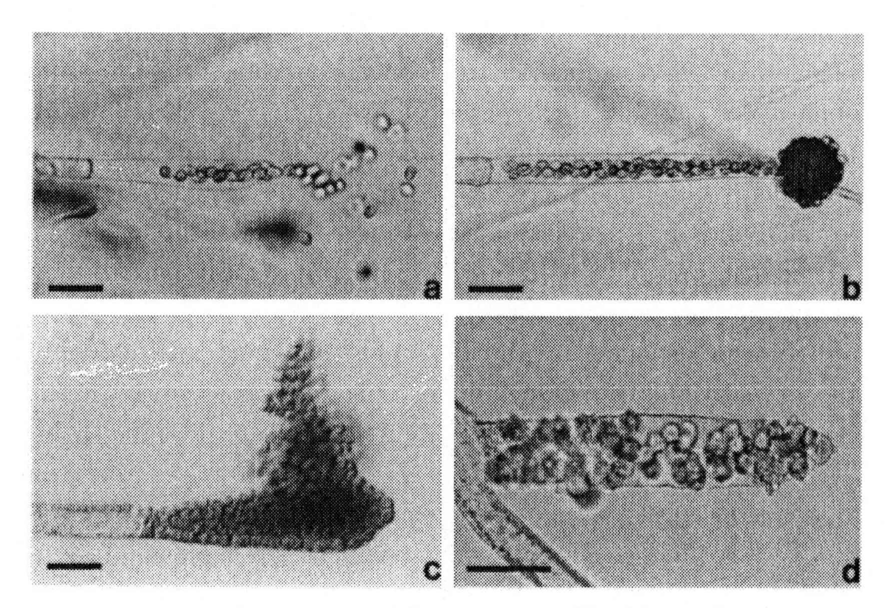

Fig. 6.5 Sporangial emptying in different water molds. (a) *Saprolegnia*: primary spores swim after ejection from sporangium. (b) *Achlya*: discharged spores cluster at apex of sporangium and become cysts. (c) *Thraustotheca*: spores spill through wall of sporangium. (d) *Dictyuchus*: cysts form inside sporangium, and each discharges a single secondary zoospore through an exit tube. From J. Daugherty, et al., *Fungal Genetics and Biology* 24, 354–363 (1998, reprinted with permission).

next 30 seconds. Even when I'm wrong (hardly ever), students look at their sporangia very carefully. As my entree to experimental mycology, Webster suggested that I try to solve the mechanism by which the spores were discharged. The process is significant for a number of reasons. In frog diseases, human pythiosis, potato blight, and all of the maladies caused by zoosporic fungi, zoospores are the agents that initiate the infection. It would seem useful therefore to understand how they are produced and expelled.

In 1923 a mycologist named William Coker said, that "without any doubt . . . the spores are expelled by internal pressure,"[10] but others invented expulsion mechanisms involving swelling gels, electrostatic forces, and webs of contractile proteins strung inside the sporangium, or simply concluded that the spores swam out under their own steam. Only Coker was correct. Breakage of the sporangial membrane and partial collapse of the cell seem to contradict this mechanism, but the pressures needed to spit a clutch of microscopic spores are very small indeed. Before their cytoplasm is cleaved, sporangia are pressurized to 4 or 5 atmospheres, just like the hyphae from which they develop. But when the membrane breaks, the pressure drops by 99.8 percent, to a value of one-hundredth of one atmosphere. A mathematical model devised by Roland Ennos, now a professor at the University of Manchester, demonstrated that this residual pressure was sufficient to push spores through the sporangial nozzle.[11]

By washing sporangia with compounds that counteract the normal flux of water into the sporangium, we found that it was possible to slow the motion of the spores, halt them in the middle of their transit, and even reverse their movement. This was discovered by accident while studying the effects of synthetic polymers called polyethylene glycols—chains of antifreeze molecules—on sporangial development. The first time I reversed spore discharge and witnessed spores backing away from the nozzle, I rubbed my eyes in disbelief, became overwhelmed by emotion, and then felt as if my head would catch fire. There are no shortcuts to this most peculiar bliss. For hundreds of millions of years, in the absence of molestation, water molds had been crafting sporangia and expelling spores. Now, for the first time ever, spores had reversed direction. Immediately, fungi with biflagellate spores became the laughingstock of every pond. Young colonies still have nightmares about the evil one whose gigantic eyes forced great-grandma's spores to retreat up her sporangium.

Polyethylene glycols proved useful for measuring the size of pores that passed through the wall of the sporangium. This information was important because we were interested in the identity of the substances that were retained in the sporangium and generated its tiny pressure. I was convinced that I had conceived an ingenious method for measuring the pores and became very proud of myself until I discovered that another scientist had arrived at precisely the same method seven years previously. When I met this biologist years later, he said that the idea came to him as a daydream when he was smoking a joint, and he imagined cells inflating and deflating as molecules wafted through their walls. He told me this story one evening in Hawaii. The next evening I was married on a beach. Not to him, I should add.

All zoospore-producing fungi use pressure to shunt their spores from sporangia, but there are many variations on this theme. *Pythium* species, such as the pathogen that causes mammalian pythiosis, form spherical sporangia that employ a two-stage emptying mechanism. Initially, the cytoplasm inside the sporangium is shifted as a single mass into a bag that inflates from a special plug of material that caps an exit tube. Cleavage of the cytoplasm into spores occurs inside this bag, and once the zoospores are mature, the bag bursts and they swim away. Species of *Phytophthora*, including the potato blight pathogen, partition their cytoplasm inside the sporangium but still shift the contents of the sporangium into a bag before they swim away. The reason that these microorganisms use the two-stage mechanism is a mystery, but a further difference between this process and the system of spore release used by *Saprolegnia* is found in the nature of the cells that swim away from the sporangia. The spores released by *Pythium* and *Phytophthora* are the kidney-shaped secondary-type zoospores rather than the pip-shaped primary spores expelled by *Saprolegnia.*

To explore this a little further it is useful to consider the fate of those primary spores of *Saprolegnia.* Upon release they swim with a counterclockwise helical path at speeds above a tenth of a millimeter a second. They then collect at the surface of the water or attach to a surface, immediately retract their flagella, and form a cyst by rounding off and secreting a cell wall. Later, the cytoplasm within the cyst empties onto the surface, resprouts flagella, and is converted into a secondary zoospore that looks identical to the swimming cells of *Pythium* and *Phytophthora* (compare Figure 6.3 b with Figure 6.3 c). Secondary spores swim twice as fast as the primary spores and follow a clockwise path. It has been suggested (and this does make a great deal of sense) that the secondary zoospore is a more recent evolutionary innovation than the slower-moving primary spore and that it was eliminated by the ancestors of *Pythium.* Based on this conjecture, the formation of the bag may be related to the developmental steps that have been edited from the life cycle in favor of a more direct leap to the generation of the secondary spores. This seems more likely than the alternative explanation, which would suggest that the primary spore is the more recently evolved vehicle for transporting oomycete genes. After all, it is not a graceful swimmer and is also missing from the life cycles of most of *Saprolegnia*'s close relatives in the family Saprolegniaceae.

Achlya is one of these relatives. For over 100 years mycologists argued about the structure of the spores shot from its sporangia (Figure 6.5 b). Many experts in the nineteenth century thought that the spores swam out of the sporangia, so the presence of flagella seemed essential to their motion, even if they could not be seen. This was the view championed by Marcus Hartog, a professor at Queen's College in Ireland, an irascible gentleman who had been a student of Anton de Bary, "the father of modern mycology," in Strasbourg.

> I have already shown that the probability is that flagella will everywhere be found when properly looked for in the escaping zoospores of *Achlya*.
>
> —M. M. Hartog, *Annals of Botany* 2, 213 (1888)

Some of Hartog's contemporaries had described the expulsion of spores lacking flagella and so he was charging them with sloppy work. He staked his reputation on the presence of flagella (how else could the spores leave the sporangium but by swimming?), and invented the terms aerotaxy and pneumatotaxy to describe the stimuli that led them to exit sporangia. But not for the first time in his career, Hartog was wrong. A century later, electron microscope views of thin slivers of sporangia revealed that the usual elongated flagella were replaced by short stumps in *Achlya* spores: all the flagellar stumps can do is wave a pair of phantom limbs.[12] If the great Marcus Hartog had lived to read about this discovery, I imagine that he would have burned with rage, hurled the journal across his office, and immediately written an article accusing the researchers of deception. He may, of course, have been a much calmer man, but if so he was a rare scientist indeed. We'll meet Hartog again in the following chapter.

As they are discharged from the sporangium, *Achlya* spores stick to one another around the outside of the nozzle and are immediately converted into cysts. *Achlya* sporangia can be recognized in ponds without a microscope. If you find the floating corpse of an insect or fish surrounded by a halo of filaments, you've located an oomycete water mold. Close inspection may reveal tiny white clubs at the tips of the hyphae, which are the sporangia. Finally, the presence of opalescent blobs at the tips of sporangia confirms the identification of *Achlya*. The blobs are the clusters of cysts from which the secondary spores emerge and rocket away from the parent colony in search of the next feast.

The life cycle of other family members is modified further, with the abolition of sporangial emptying through a nozzle in favor of discharge through a wider wall fracture (Figure 6.5 c), or the formation of cysts inside the sporangium and release of secondary spores through multiple exit tubes (Figure 6.5 d). I am intrigued by the possibility that today's water molds offer glimpses of an evolutionary transition series. A familiar example of such a succession is the trip from tiny monkey through chimpanzee, a hominid with an uncanny resemblance to an NFL lineman, to evolution's Caucasian climax with his (never her) purposeful stride toward the office.[13] Although transition series can convey useful messages about the kinds of changes that must have occurred during the evolution of particular organisms, many students become confused by the idea that there is an inexorable progression toward some particular species. For instance, a friend of mine in high school once divulged that he couldn't make sense of evolution because if living things had been around for so long, everything should be human by now. For this reason it is unforgivable to arrange contemporary species in a scheme of ever-increasing complexity, but some biology textbooks continue to encourage these misconceptions.

Returning to water molds, there is a logical transition that probably links different types of emptying mechanism found in the Saprolegniaceae. This would begin with something like today's *Saprolegnia,* which spits motile primary spores from its sporangia, through *Achlya,* which has purged the first swimming stage but still empties all of the spores from its sporangia through a nozzle, to other water molds, which have dispensed with the graceful mechanism of sporangial emptying in favor of cyst formation inside the sporangium followed by the discharge of secondary spores. Because the primary spores are poor swimmers it is easy to believe that their disappearance confers some selective advantage, but this does not explain the fact that business is thriving for *Saprolegnia* and its ungainly life cycle. Some molecular genetic evidence suggests that *Saprolegnia* has been around much longer than other members of its family, which may be consistent with the idea that its mechanism of emptying sporangia is primitive and served as the predecessor of processes adopted by other water molds.

The genetic modifications necessary to convert water molds from one mode of spore release to another may be surprisingly simple. I say surprising, because the sporangia look very different from one another as they empty (Figure 6.5). For instance, to convert a *Saprolegnia*-type sporangium into an *Achlya*-type sporangium would only require mutations (or perhaps a single mutation) that prevented elongation of the flagella. Sporangial development involves a series of discrete processes, with the completion of one step triggering the next. In an engaging book published in 1979, Ian Ross suggested that after cleavage, spore release requires the synthesis of a single enzyme that weakens the wall of the nozzle.[14] If Ross is correct about that nozzle enzyme, its absence would disrupt the emptying process seen in *Saprolegnia* and *Achlya*, resulting in cyst formation within the sporangium as we see in some of their relatives. In most respects, the appearance of the sporangia of different water molds is almost identical until just before the spores are expelled. Details of the final steps in sporangial development almost certainly determine everything about the mechanism of emptying and these all relate to patterns of enzyme secretion.

When water is absorbed by hyphae, their cytoplasm expands and pushes the plasma membrane against the inner surface of the cell wall. In this way, fungal mycelia become pressurized. Water also flows into zoospores as they glide through the water, but they are not protected by a cell wall and will burst unless they expel excess water. This osmotic adjustment is controlled by a structure called the contractile vacuole, which squirts water from the spore at a feverish rate of once every 5 or 6 seconds. When zoospores locate a suitable surface, they anchor themselves by spitting a glycoprotein glue from tiny vesicles under their plasma membranes. The primary spores retract their flagella and presumably use their components to fashion new flagella as the cell is rearranged into the secondary form. When secondary spores stop swimming they eject their flagella into the water and this dismissal of their motors marks the microbe's entry into the mycelial phase. As the flagella are cast off, the spores stop pumping water, immediately puff up into spherical cysts, and quickly assemble a protective wall on their surface that prevents bursting. Once pressurized by water uptake, the spore acts as a drilling rig and pushes a slender hypha into the underlying surface. In pythiosis, cysts germinate in skin wounds, and other oomycetes burrow into amphibian skin, the necks of turtles, between fish scales, on the surface of roots, and any scrap of plant or animal debris that they can dissolve. The region of the cyst surface from which the hypha develops is not chosen at random, but must be on the same side as the point of attachment to the food. In some species, the zoospore settles with its groove oriented toward the landing site, and it is from this predetermined spot that the hypha extends.

Zoospores are chemotactic, meaning that they can sense chemical gradients and use them to home in on particular targets. In the simplest case, sugars and amino acids exuding from wounds on plant roots signal that food is nearby, and seepage from animal tissues may offer zoospores the same kind of lure. Electrical fields provide more subtle cues for water molds that infect plants. All eukaryote cells disclose themselves with weak electrical currents that result from the activity of ion pumps. The response to these signals is called galvanotaxis. The signature of healthy roots differs from that of wounded ones, and zoospores respond to one or the other cue depending upon their preferred quarry. In the laboratory they can be enticed by electrodes delivering the same strength of electrical field as a root, providing the investigator with a Pied Piper experience similar to my control over spore motion inside the sporangium.

The zoospore is far more complex than an Ingoldian conidium. A conidium is a passive vehicle for transmission and has no ability to counter environmental challenges. If it doesn't hit a submerged leaf, it is eaten by a predator, suffocates in mud, or is washed downstream to oblivion. Zoospores are elegant swimmers and are equipped with onboard targeting hardware that enables each cell to respond to its surroundings. They can spend hours exploring a small volume of wet soil, navigating around obstacles and examining surfaces for the optimum landing pad, before settling down and initiating invasion. The spores of some chytrids will swim toward a surface, crawl over its hills and valleys, and make a cyst if they are satisfied or swim away in search of more fertile territory. Confined to a drop of water on a microscope slide, all zoospores manifest a determination to survive. They avoid the heat from the microscope lamp by swimming around the edges of the drop where the water is coolest, and bounce off the glass as they hunt for an escape route. Their galvanotactic sensing mechanism is probably overwhelmed by the electrical equipment in a modern laboratory but they soldier on, searching for a nutritious landing spot until they have exhausted their cytoplasmic fuel supply. Even then, they'll form a cyst in hopes of a better future. It is their motility that allows me to empathize with their struggles, but at a less obvious level, all cells manifest the same vitality. Any biologist who has looked at individual cells with a microscope and poked them with needles knows that none of them embrace death. They all squirm and try to seal their leaking membranes. (I'm glad my vivisection is limited to the exploitation of fungi; I'd commit suicide before I took a scalpel to a kitten.)

On a final note, I find it fascinating that two entirely dissimilar modes of existence are displayed in the life cycle of a single oomycete water mold. Zoospores swim and pump water from their cytoplasm; hyphae soak up water and use the resulting pressure to push through tough obstacles. All fungi, mushroom relatives and oomycetes alike, are believed to have evolved from ancestors that swam with flagella, and it is very likely that swimming spores are more ancient inventions than branched mycelia. In a sense, then, the water mold displays a primeval fragment of its history every time it stops extending hyphae and ejaculates a mist of zoospores into the water.

CHAPTER 7

Siren Songs

> The high, thrilling song of the Sirens will transfix him,
> lolling there in their meadow, round them heaps of corpses,
> rotting away, rags of skin shriveling on their bones.
>
> —Homer, *Odyssey*

Bewitched by the sirens' song of Greek mythology, mariners were enticed to their deaths on the jagged rocks of the Straits of Messina. Before Odysseus's voyage, the goddess Circe enlightened him about the behavior of these avian seductresses. She advised him to order the crew to plug their ears with wax and have himself lashed to the ship's mast if he wanted to savor the song but frustrate access to the rudder. The Icelandic diva Björk can distract me from almost anything, sweep me to despair or to delight with a few bars of her angelic warbles, but I'm not convinced that she could compel anyone to commit suicide. This persuades me that a song capable of damning a ship's crew can only have been perceived as an advertisement for sensational sexual favors. Only carnal desire has the power to make a hero act like a fool.

Songs are sung within the human body by molecules rather than tuneful pressure waves, but they have an equally inescapable allure. Consider the behavior of spermatozoa, unicellular vehicles for half our genome, built to swim up the scented gradient wafting from a ten-thousandfold larger egg. Likewise the fungi. They sing to one another in chemical pulses, and answer with a symphony of developmental changes. This chapter explores the mechanisms that enable fungi to communicate with one another during reproduction—Where are you? Who are you?—and related signaling processes that allow hyphae to cooperate when they form fruiting bodies.

Oomycete water molds produce eggs in sacs called oogonia that swell at the tips of hyphal branches (Figure 7.1). Male structures called

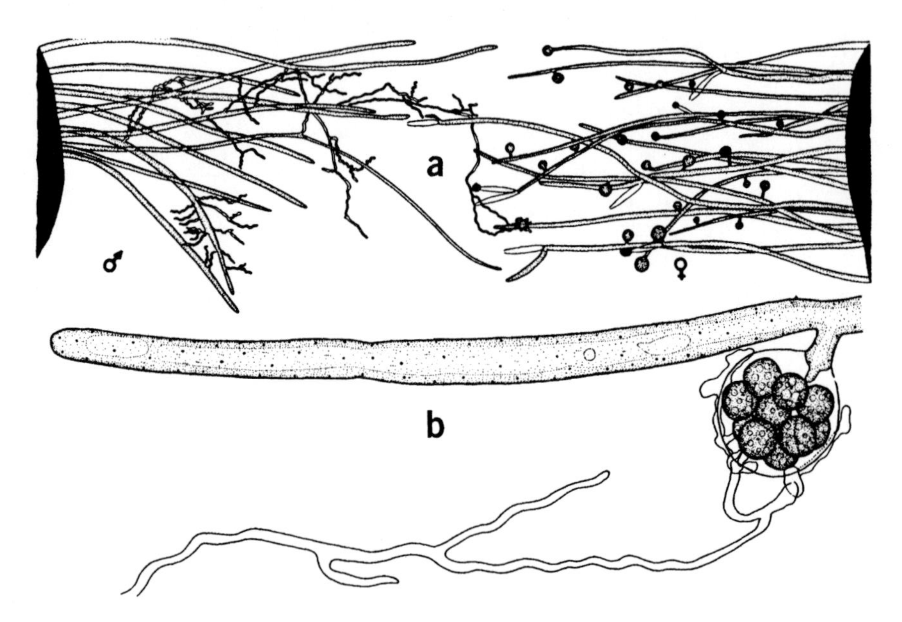

Fig. 7.1 Sexual reproduction in the water mold *Achlya ambisexualis.* (a) Hyphae of male and female strains growing from hemp seeds. The male has formed antheridial branches and the female has formed oogonia. (b) Higher magnification view of contact between antheridium and oogonium. Note that this antheridium has penetrated the surface of the oogonium at multiple locations. After fertilization, each of the oospheres will become a thick-walled oospore. From J. Webster, *Introduction to Fungi* (Cambridge: Cambridge University Press, 1980), reprinted with permission.

antheridia also develop as offshoots from hyphae. These slender growths attach to the surface of the oogonia, penetrate the eggs (called oospores after fertilization), and inject sperm. Water molds and other fungi advertise their readiness for intimacy using molecules that have been termed hormones, pheromones, and even morphogens. Pheromone seems the most appropriate noun, because hormone and morphogen are commonly used to describe compounds generated by larger multicellular organisms that affect cells within their own bodies. The pheromones of water molds are called antheridiol and oogoniol. The chemical structures of antheridiol and oogoniol are very similar to the human sex hormones estrogen, progesterone, and testosterone, which is remarkable given the remoteness of the evolutionary relationship between water molds and animals. All of these compounds are ring-shaped lipids called steroids, which are employed for relaying messages between all kinds of cells.

We know very little about the ecology of water molds, but it seems that they have sex in some remarkably unpleasant places, often in stagnant ponds attached to dead insects or fragments of decomposing leaves. In the pond's nutrient-rich soup, the female makes the first move, releasing antheridiol from the lacework of her mycelium. If a male is growing sufficiently close, he responds by producing his antheridia. As these grow toward the source of the antheridiol they release the second pheromone, oogoniol, which is recognized by the female. Sensing the success of her seductive behavior, the female forms oogonia or egg sacs, and then secretes more antheridiol to inflame every red-cytoplasmed male in the neighborhood. She soon becomes entangled in a cage of male sex branches that clamber around until they pinpoint the source of the antheridiol trailing into the water. Once contact is made with the oogonium, the tips of the antheridia flatten against its curved surface and this connection stimulates egg formation. Finally, the egg chamber is penetrated by stumpy hyphae that protrude from each antheridium, and the successful male injects sperm nuclei into one or more eggs. The aquatic liaison is consummated. When multiple males reach the oogonium at the same time, each of the eggs in a single clutch can be fertilized by a different male. After fertilization, the eggs develop thick walls and remain dormant until environmental conditions favor germination and growth of a new mycelium.

John Raper, who was at Harvard, was the first to demonstrate that the development of antheridia and oogonia was coordinated by chemical signals.[1] He worked with different strains of *Achlya ambisexualis* (which he discovered), and its relative *Achlya bisexualis*, isolated from ponds. Fishing for oomycetes requires finesse. Samples of pond water and sediment are brought back to the laboratory and poured into glass dishes. Sterilized seeds are then sprinkled onto the surface of the water to summon zoospores from the sediment on the bottom of the dish. Activated from their cysts, the motile spores swim toward the water surface, glue themselves to the seeds, and penetrate their shells. Within a day or so, each seed sprouts a fringe of water mold hyphae. *Cannabis* seeds have a very high fat content and are superb baits for water molds. I have no idea how this discovery was made, but in my lab we keep bags of *Cannabis* seeds. They are grown in greenhouses for scientific purposes and shipped (to my students' dismay) presterilized and incapable of germination. Following a painstaking process of purification, cultures of single strains are obtained from the buoyant colonies and can be maintained indefinitely on agar. Some strains have been tormented in laboratories for decades—snatched from their ponds and given no choice but to extend their hyphae and perform sex acts with partners selected by spectacled academics. Invariably, the reward for their compliance is euthanasia in the pressurized steam of an autoclave.

Oomycetes secrete two types of antheridiol to attract males, and a series of different oogoniols to control female behavior. In animals, steroid hormones enter the target cells and bind to receptors in their nuclei. The hormone-receptor complex then acts as a transcriptional regulator, directing the expression of specific genes that control the response to the hormone. This mechanism has not been confirmed in *Achlya*, but seems likely. Physiological experiments provide insights into the male response to antheridiol and give hints about the frantic gene expression in the male hyphae once they become aroused by the proximity of a female. In contrast to the usual hyphal branches that grow some distance behind the tip (see Figure 3.1 b), antheridia begin life as bumps that emerge as extensions from the cell wall close to the apex (Figure 7.1). When the antheridiol message is received, the hypha stops extending and growth is refocused to these spots just behind the tip. These observations are consistent with the idea that the antheridia are sculpted by localized changes in the plasticity of the hyphal wall. Evidence to support this model was found in 1967 by a student, Donovon des Thomas, and his mentor John Mullins, at the University of Florida.[2] They discovered that the appearance of male branches coincided with the secretion of a type of enzyme called an endoglucanase. Since the female hormone antheridiol elicited both a surge in enzyme activity and branch formation, Thomas and Mullins suggested that antheridiol induced endoglucanase secretion, and that the enzyme clipped polymers within the wall, causing it to relax.

Evidence that endoglucanases actually cause wall loosening did not emerge until thirty years after the publication of the Thomas and Mullins paper. Terry Hill is a professor at Rhodes College in Memphis. In the 1970s he had been a student in Mullins' lab (after serving as a gunner on a river boat that motored up and down the Mekong River during the Vietnam War—a rare experience for a mycologist). During discussions at a conference, Terry and I realized that we could test the wall-loosening idea by combining a couple of techniques. At the time I had discovered that oomycetes cultured in solutions containing high concentrations of sugars lost their internal turgor pressure and made a more plastic cell wall. If this behavior was pushed to an extreme, the wall became very fluid, causing the hyphae to abandon their normal cylindrical shape and splurge across the agar surface like jellyfish stranded by the tide. The rationale for the new experiment was simple. We would test whether wall strength and enzyme activity were associated by measuring both variables from cells cultured in a range of sugar concentrations to induce wall loosening. A few weeks later, Terry measured endoglucanase secretion from hyphae while I determined the strength of their walls.

The examination of wall strength was tricky, requiring the use of a complicated experimental rig. First, a glass micropipet filled with a transparent oil was inserted into a growing hypha with the aid of a micromanipulator. After impalement, the pressure inside the pipet was raised so that oil was injected into the cell. This compressed the cytoplasm and stretched the hyphal wall. Then the pressure inside the pipet was increased further until the wall ruptured and oil spurted over the agar (Figure 7.2). The tensile strength of the wall was calculated from the critical pressure recorded an instant before the cell burst open. Afterward, the normally animated hyphae were transformed into transparent, lifeless tubes that continued to spew oil. We were delighted. There was a strong correlation between the activity of the enzymes secreted by the hyphae and the strength of their walls. High enzyme activity was associated with more plastic walls (Figure 7.3). While this didn't prove a causal link, it did support the conclusion of the Thomas and Mullins paper.[3] Endoglucanase secretion is associated with wall loosening and this fundamental mechanism may govern the emergence of male and female branches.

Surprisingly, my paper with Terry Hill has not stimulated a telephone call from Sweden, but the scientific elite was not always so dismissive of

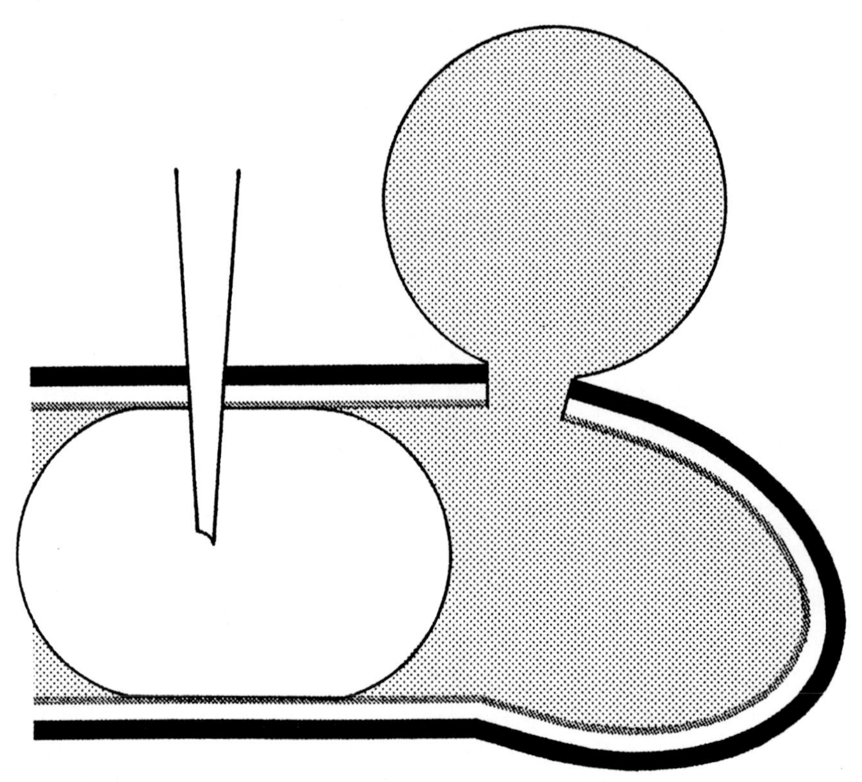

Fig. 7.2 Technique for rupturing cell by pressure-injection of silicon oil for measurement of cell wall strength. Method described in text.

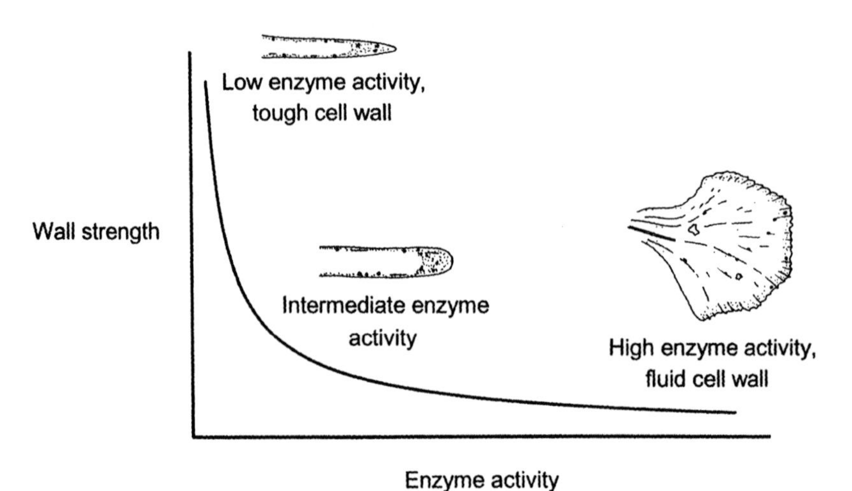

Fig. 7.3 Graph showing relationship between activity of secreted endoglucanases and the tensile strength of the hyphal wall. Drawings show the form of cells with different levels of wall strength.

water molds. Oomycete sex was once considered one of the most important botanical mysteries, and bitter professional rivalries developed among the scientists engaged in its study. Marcus Hartog considered himself an expert on the subject. He had been a student of Anton de Bary, "the founder of modern mycology," at Strasbourg. Later, as a professor at Queen's College in Ireland, Marcus assaulted the work of his competitors in published commentaries on fungal reproduction. For example, he took a particular dislike to a Professor Trow at University College in Cardiff:

> I must express my regret that so careful a technique and such powers of observation as the author's [Trow's] should have been overweighted with a priori and obsolete views. It is not good to follow too literally the adage 'Eher eine schlechte Hypothese als gar keine.'
>
> —M. M. Hartog, *Annals of Botany* 10, 100 (1896)

My friend Barbara Nemeth has translated the maxim as, "It's better to have a bad hypothesis than none at all." Hardly the signature of a sociopath, but the Irish professor's ravings contrast greatly with the politeness of scientific discourse among mycologists today. Hartog's ire rose when Trow reported that water mold eggs were fertilized by male antheridia (precisely as I have described). Hartog was unmoved by the evidence and referred to an earlier quip by de Bary that fertilization in these fungi was an illusion that could be likened "to the passage of spiritual mediums like the renowned Mrs. Guppy through brick-walls and closed doors and windows." But just as he had been mistaken about the zoospores of *Achlya* (Chapter 6), Hartog severely underestimated the quality of Trow's work on oomycete sex. Within a few years, Trow was completely vindicated when other mycologists corroborated his version of events.

Some historical context may help in evaluating the source of Hartog's frustration. Trow was using a technique called serial sectioning, which Hartog regarded with suspicion. Trow pickled fungi in an acidic solution (or fixative), embedded them in paraffin wax, and then cut shavings (a few thousandths of a millimeter thick) from the specimen using an instrument called a microtome, fitted with a razor-sharp steel blade.

Before viewing, the slivers were stained with gentian violet, eosin, and other dyes to enhance the contrast between different cell components. In the hands of a master microscopist, this was a powerful method. By studying successive slices from single oogonia Trow was able to reconstruct their three-dimensional structure. This enabled him to determine the location of nuclei within the intact cells and to piece together a logical progression of the events that led to egg formation. The task was especially difficult because sperm nuclei are produced inside the antheridium just before they are propelled into the egg, making it almost impossible to obtain a specimen in which fertilization was taking place at the time it was paralyzed in fixative. Trow's success with the method enraged Hartog, who clung to his practice of looking at whole cells crushed on a microscope slide.

Although most experiments on oomycete sex have been performed on so-called heterothallic species that require a mate for egg production, the majority of water molds are self-fertile hermaphrodites that produce antheridia and fertilize their own eggs. Some species have dispensed with sperm entirely and form oospores without the appearance of any antheridia. This is called parthenogenesis. Clues to the importance of egg formation—even when it happens without sex—have come from recent work by Don Thomas, long since transplanted from Florida and now an elder statesman at the University of Windsor in Ontario. Don has compared the distribution of sexes of *Achlya ambisexualis* in seasonal ponds that dry out in summer and freeze solid in winter, with deeper ponds that remain fed by spring water year round and never freeze completely. *Achlya ambisexualis* is an unusual species because it incorporates a family of strains in which distinct males and females coexist with hermaphrodites, and she-males that offer oogonia or antheridia according to the gender of available partners. In the seasonal ponds, there are 10 times more self-fertile hermaphrodites than mate-requiring heterothallic strains; in the permanent ponds, heterothallic strains are far more common. From these observations it seems likely that self-fertility is a particularly effective strategy for a microorganism that lives in an unstable environment. The environmental stress in this type of habitat places a premium upon fast egg production when water is abundant, rather than waiting for an encounter with a stranger of the opposite sex (or one of those accommodating she-males). At least in the short term, the production of eggs outweighs the potential benefit of genetic recombination, even though this procedure smothers genetic innovation within the population. The primary advantage of egg formation in the seasonal pond may be the provision of a survival capsule, because the thick-walled eggs allow *Achlya* to persist in the drying mud like a lungfish. This may explain why the egg-laying behavior has been retained by hermaphrodites and parthenogenetic species that probably evolved from heterothallic water molds.[4]

I learned about Don's work a few years ago when he insisted on taking me to see his ponds before I gave an afternoon seminar in his department. Arriving in a suit and tie, I was poorly dressed for a fungal foray. What would anyone but a scientist have concluded after seeing two men in the woods, one in rubber boots, the other prancing between the bushes to protect a very handsome Italian jacket, both talking heatedly about hermaphrodites? John Webster tells a wonderful story about an encounter with a farmer who caught him collecting bagfuls of cow manure and rabbit droppings. He attempted to explain that he was isolating coprophilous or dung-loving fungi, but remaining astounded at this perversity, the farmer telephoned the police.

When making sense of life on earth, biologists (and filmmakers) usually rely on images of big cats and bleeding gazelles and ignore things like water molds that are hidden underwater in their parallel lilliputian universe. But to this mycologist, especially after the first cup of coffee in the morning, a backyard pond can seem as red in tooth and claw as the Serengeti Plain, and far more inspiring. When the perfume from a drowning caterpillar disperses in the water, swarms of zoospores streak toward the writhing insect. Once the spores strike their target, they glue themselves to the exoskeleton, and bore through its layers of chitin with hyphae that proceed to pulp the creature's tissues. After thorough excavation of the bloated corpse, the nourished mycelium bursts from its surface, converts its hyphal tips into sporangia, and spits the next generation of zoospores into the water. And once in a while, mycelia woo their mates with steroid songs and lay some microscopic eggs. (The plastic-lined pond in my yard deserves National Park status. It even has a frog.) These microorganisms illustrate the dazzling logic of evolution's central principles as clearly as any mammal. The continuing presence of water molds demonstrates that individuals of the various species have established a perfect track record for producing offspring. Those organisms that function well enough to transmit their genes do so. What is living today has worked; whatever works stays. Lion genes make good lions, worm genes make good worms, and *Achlya* genes make great water molds.

Other fungi enjoy equally inspiring sex lives. Like their frog-killing zoospores described in the previous chapter, the eggs and sperm cells of chytrids are motorized with a flagellum. These motile gametes synthesize pheromones from compounds called sesquiterpenes rather than steroids. The chemical released by the male sex cell is called parisin, and the female chemical, so aptly, sirenin. When sperm cells detect sirenin, their behavior changes abruptly. They stop wandering around in the water and immediately begin swimming in a corkscrew path toward the egg. To avoid overstimulation, the sperm cell inactivates sirenin molecules as it swims through the pheromonal cloud. This allows it to take frequent directional bearings from the concentration gradient originating from the egg, a crucial activity for a cell that is likely to be competing with many other gametes. When the sperm cell gets close to the egg, it approaches in a series of short dashes until it collides with her surface. The parallel between the behavior of the chytrid and siren-besotted mariners is provocative. Despite occasional signs of gentility, the microbial essence of human behavior is unmistakable.

Zygomycete sex is showcased in every textbook of biology as an example of the disgraceful antics of lower life forms, preceding the much lengthier treatment of animal reproduction. Even high school students are familiar with the warty, black zygospores of zygomycetes, although they spend far more time defiling textbook diagrams of human anatomy than understanding fungi. (The garnished illustration of the human skeleton in my biology book would have embarrassed the Marquis de Sade.) There is no common name for the zygomycete fungi, although "pin mold" is a term that was once used to describe those that grow on bread. Each pin is a stalk topped with a bulbous spore-filled sporangium (see Figure 7.4). These were the first fungal structures described with the aid of a microscope. Robert Hooke found them growing from a

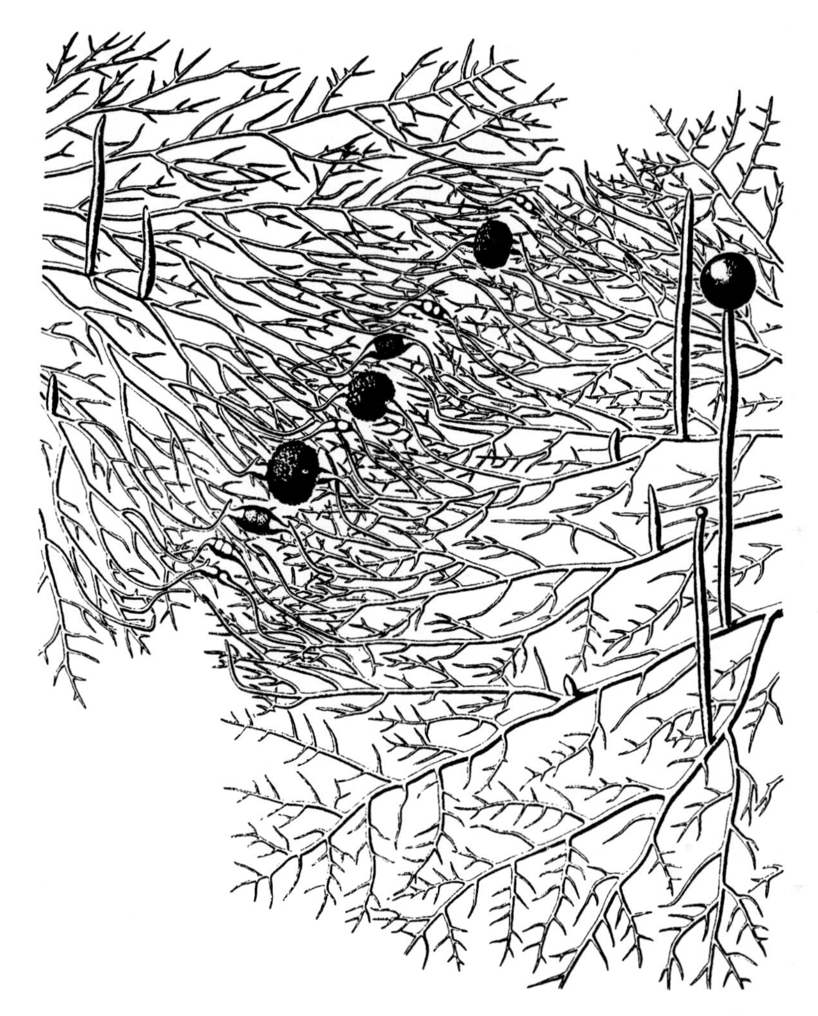

Fig. 7.4 Mating between two strains of the zygomycete *Mucor mucedo*. Asexual sporangia (tat, upright branches) developing on both mycelia; sexual zygosporangia forming only in central contact region. Different stages in zygosporangium formation are illustrated in the contact region. From A. H. R. Buller, *Researches on Fungi*, vol. 4 (London: Longmans, Green, 1931).

sheepskin book cover and illustrated a forest of sporangia in his *Micrographia* of 1665. (Delight your family by moistening a bread slice, leaving it on your kitchen counter for a few days, and providing a magnifying lens for viewing the resulting thicket.) The spores inside the sporangia are created by repeated mitotic division of nuclei shuttled from the mycelium. This is an act of asexual reproduction. Life on a bread slice begins when one of these spores drops from the air and alights on the rough terrain. If the bread is moist, the spore germinates and sends out the first thin hypha of a mycelium that will eventually colonize the entire slice. The mycelium penetrates the bread by softening it with enzymes and thrusting forward with its hyphal tips. Enzymes ease the passage of the mycelium (although bread offers little resistance) and also supply it with readily absorbed sugars. When the fungus breaches the surface and encounters air again, it enters a distinct developmental pathway to abandon the depleted substrate: hyphae send up aerial branches and their tips are transformed into the sporangia. With various ornaments, the life cycle of every fungus pursues a similarly endless progression of spores, mycelium, spores, mycelium. Using this strategy, a zygomycete clone from a single spore might thrive on kitchen scraps for the entire life of a house. But when two strains meet in a sodden slab of pizza crust, they embrace the opportunity for gene shuffling.

The term zygomycete refers to the zygospore, the spore produced between two compatible strains or on one contortionist hermaphrodite that fertilizes itself. There are about 900 species of zygomycete fungi, whose zygospores range from colorless spheres to the jet-black nuggets covered with warts shown in textbooks. These spores are similar to water mold eggs. Both are lethargic babies, adapted for endurance without food and water and possessing none of the attributes associated with spores designed for dispersal. Melanin and other pigments render the wall surrounding the zygospore[5] impervious to damage by ultraviolet light, noxious chemicals, other microorganisms, and even heat. Zygospores formed in rotting food become buried in landfills, and those that missed the charity offered by a dirty kitchen and dined al fresco sleep in the soil.

Once again, the difficulty of sex for zygomycetes is recognition. With whom should I attempt to fuse today? Zygomycete pheromones are manufactured from carotenoids, the class of compounds that includes ß-carotene, which is the precursor of vitamin A, and pigments that function in plant photosynthesis. Rather than releasing great quantities of these chemicals into their environment, zygomycetes have evolved a more economical scheme for signaling. Each strain releases a minute amount of its own precursor compound and this is converted into a pheromone called trisporic acid upon receipt by the partner strain. An individual strain cannot process its own precursor into trisporic acid; only its mate possesses the necessary enzyme. This type of chemical interaction is called cooperative biosynthesis. In response to the accumulation of trisporic acid, formation of asexual sporangia ceases and both strains produce hyphal branches that extend into the air and grow toward one another. Like the oomycete pheromones antheridiol and oogoniol, trisporic acid and its precursors function in water, but they are also transmitted through air, promoting a cobra-like dance between the aerial hyphae that climaxes in fusion and zygospore generation.

Sexual behavior in mushroom-forming fungi spans monogamy and civility, to group sex and slaughter. Mushrooms develop after compatible strains fuse in the soil. We don't recognize different sexes in basidiomycetes because all mycelia appear as masses of androgynous tubes. But genetic diversity lurks in the nuclei housed in these hyphae. A single species of ink-cap mushroom can encompass hundreds of different strains, so in the sense that "male" and "female" refer to compatible mating types (as they do in humans) there are legions of different sexes of basidiomycete fungi. A handful of genes determines whether individual mycelia can fuse and go on to raise a family of spores on the gills of a mushroom. Each of these genes can exist in different versions called alleles, but only certain combinations of alleles can cohabit a single mycelium. When incompatible strains meet and their hyphae fuse, love is lost in the cytoplasm and the region of mingled juice darkens and decays. The percentage of failed mushroom marriages is probably very high. But the evolution of so many sexes makes sense because lots of combinations can work together and this outcrossing favors the perpetuation of tremendous genetic variation in a single species. Presumably, this variation translates into physiological flexibility when we consider a population of ink-cap mycelia. Ink-caps and ink-cap genes have certainly been around long enough to testify to the success of this reproductive strategy.

While conjugal strife between ink-caps is hidden underground, any patch of woodland offers a showcase for discord among other types of fungi. If your knowledge of Balkan geography is as poor as mine, I think I could convince you that a map of this tormented region is etched in black ink on tree stumps and on the ends of logs. Fungi define the borders in the wood, but Balkan history offers a good analogy for the origin of the inscriptions. Beech trees are the battlegrounds for swarms of wood decay fungi in Midwestern forests. Older trees with fractured crowns are shot through with hyphae and sprout clusters of brackets as large as park benches. The destruction of a big tree can take a century or more and preoccupies a rich biological community. To breach the bark, fungi rely upon wood-boring insects, squirrels, sapsuckers and woodpeckers, and even lovelorn morons with penknives. Once these vandals have acted, an ancient mechanism is initiated. Somewhere else in the woods a drop of liquid forms at the base of a basidiospore, and the propagule is catapulted from its basidium. Two seconds later gravity pulls the spore clear from the bracket and it is swept away in the wind. One among billions, this spore lands in a hole drilled by a sapsucker, or in the crotch of the *v* of an "I love Mary," and the infection begins. It is a damp afternoon. The spore absorbs water and germinates. First one hypha, then a fan of branching filaments, and the fungus works its way deeper into the trunk. Year after year, secreted enzymes convert wood into sugar molecules and the sugar is absorbed by the fungus and burned in its mitochondria. But the young mycelium is not dining alone. Other holes were dug through the bark, the woodland air is filled with spores, and numerous fungi feed on every tree. When these mycelia meet, they fight or fuse. Conflict draws the Balkan map. Where different species and incompatible strains of the same species clash, the wood becomes stained with pigment as the warring mycelia attack their opponent's hyphae and produce thick, melanin-impregnated walls to resist each other's poisons. So when the tree is cut down or blown over, its trunk displays a map that outlines the borders between microbial rivals. Each country on the map is actually a three-dimensional column, the size of a baseball bat or longer, that runs up and down the trunk. Compatible strains fuse and leave no black line between them in the wood. The only obvious sign of these successful unions are the brackets that grow outside the tree.

Alan Rayner at the University of Bath was the first to appreciate the message of diversity offered by decay columns.[6] By isolating fungi from different sectors of the decomposing wood, pairing them in different combinations on agar, and becoming a voyeur to their gladiatorial clashes or acts of microscopic sex, Alan was able to determine how many strains were vying for the resources in a single dying tree. Tree decay is a slow business, and wood retains much of its strength even when it is stained with decay columns. Wood from such trees is referred to as "spalted" and can be chiseled into beautiful bowls and other ornaments. On my office windowsill I have a colony of model mushrooms turned from spalted beechwood by the British artist Gordon Peel. The wood came from trees felled by hurricane-force winds in England in the 1980s. Stands of beeches were severely damaged by these storms because beeches have shallow roots and are poorly buffered against high winds. Their fungal infections probably did little to hasten their ruin. Many tree species can become hollowed out by fungi and insects, leaving tubes of living wood that are sometimes better able to survive storm damage than heavier, solid trunks that are free from infection. Like other cylinders, pound for pound much of the strength of a tree lies just beneath its circumference.

Contrary to the rule that mushroom sex involves two strains, some fruiting bodies develop from more immoderate matings and contain nuclei from many strains. One such sexual extravaganza was unearthed recently in Massachusetts, by researchers studying *Armillaria gallica*. (This was the species whose ancient and hefty mycelia were discussed in Chapter 3.) The mycologists in New England had collected and preserved fruiting bodies from the same site for more than twenty years and were able to discriminate between hyphae of different strains through analysis of their DNA sequences and enzymes.[7] The evidence suggested that in some cases single mushrooms developed from nine or more individual mycelia. To mycologists used to thinking about sedate marriages between two consenting fungi, this is quite unsettling. How many of the spore-producing platforms that we recognize in the woods are communities of microorganisms absorbed in group sex? Aunt Etty would have dropped her stinkhorn stick and raced for home had she known about this sylvan debauchery.

The sexual appetite of *Armillaria* is quite erratic. While the Massachusetts fruiting bodies collected before 1988 were genetic mosaics, mushrooms that have emerged more recently in the same locations appear to develop from single mycelia. The reason for the flurry of multiple fusions followed by a period of more modest sexual encounters is unknown, but it is possible that variations in climate are responsible. In years when water and soil nutrients are abundant, fungal mycelia thrive, increasing the probability that a mycelium formed from two strains can accumulate enough biomass to fashion its own crop of mushrooms. During less favorable years, recruitment of many mycelia might be necessary to support the formation of fruiting bodies. This may be a rare case in which environmental stress encourages cooperation between individuals of the same species. But it is also possible that the apparent teamwork is deceptive. Only those strains whose hyphae emerge at the surface of the gills and produce spores will benefit directly from the alliance. This caution is recommended by the story of slime mold strains called "cheaters." Although the handful of slime mold experts working today were all trained as mycologists, slime molds are unrelated to fungi. When starved, slime mold cells aggregate and create microscopic stacks of cells that act as fruiting bodies. These can form as genetic mosaics, and in such instances different strains of the microorganism jostle for position within the developing fruiting body. Cheater strains migrate to the tip of the stack, where their cells are converted into spores, while less aggressive strains form the stalk that supports them. Few of the non-cheater cells contribute to the spore mass. They are the microbial equivalent of stepparents.

Mycologists have been bewildered by mushroom formation for decades, which explains why most have chosen to ignore it as a research subject. Even a casual inspection of a fruiting body is enough to frighten a biologist who dreams of solving something in his or her lifetime. Consider the fly agaric, the choice of fairy tale illustrators. Its bright red cap, splattered with white scales, can enlarge to the diameter of a dinner plate. Underneath, hundreds of white gills are spaced as evenly as spokes on a bicycle wheel and all of their edges hang flush with the bottom of the cap. The gills could not be more arranged more flawlessly if they were cut with laser-guided pruners. Natural selection cannot excuse small errors in the vertical orientation of the gills or in their spacing because these would prevent spore dispersal (and any mushroom that doesn't release spores would be a very sad excuse for a mushroom). Although the size of the stalk varies among individuals, its thickness is always attuned to the weight of the cap. This integrity of form signifies that hyphae in the fruiting body are articulate in communicating positional information with their neighbors. The species specificity of the resulting mushrooms informs us that different songs are sung beneath the caps of fly agarics, ink-caps, and chanterelles, but for now, science is deaf to the chemical language of these arias.

Although modern research on developmental biology is focused at the molecular level of inquiry, most experiments on development rely upon detailed anatomical studies of plants and animals. For instance, there is a published atlas that shows the position of every one of the 959 cells in the body of the roundworm *Caenorhabditis (seen-o-rab-die-tiss) elegans.* This worm has been adopted as a model for research on animal development because, among other reasons, all worms of a particular age have the same number of cells, and because 959 is not an overwhelmingly large number (e.g., the frog in my pond has 16 million brain cells). By documenting every cell division that converts a fertilized egg into an adult worm, researchers have determined the developmental origin of every cell. This anatomical blueprint is a tremendous resource for scientists trying to understand how the gut of the worm develops, the maturation of the animal's reproductive organs, or the formation of its nervous system. But the same path of inquiry has not been productive for mycologists. Unlike a worm or a plant, in which numerous types of tissue are built from distinct kinds of cells, fruiting bodies consist of nothing but tangled masses of identical cells.

At first glance through the eyepieces of a microscope, a mushroom slice appears like pavement made up of roughly spherical cells that flatten wherever they make contact. An apple or potato looks very similar under the microscope. But this resemblance is very superficial. Apples and potatoes grow by repeated division of specialized groups of cells called meristems. Meristems situated at different positions within a plant throw off cells that form the tissues of leaves, stems, roots, flowers, and fruits. Fungi do not have meristems. Instead, mushrooms are produced by the elongation and branching of innumerable hyphae. The reason that the tissues of apples and mushrooms look so similar is that it is impossible to cut along a hypha very far before passing through the other side. I'll explain this with an analogy. Imagine you have tied twenty or more rattlesnakes side by side with some strong cord. Now, assuming you could maneuver your bundle of serpents to a workbench, consider how difficult it would be to cut any individual reptile lengthwise using a bandsaw. Collapsing next to the cut ends of the bundle, your last sight would be a collection of bloody ellipses and circles. Pull yourself together; this was a useful thought experiment. A fruiting body can contain millions of hyphae, and when its stalk or cap is cut, you'll always see ellipses and circles rather than cylinders. This hyphal construction is a unique feature of fungi. Nothing else develops in the same fashion.

The first sign of fruiting body genesis is a cluster of hyphae that converge as a white pellet the size of a pinhead. This is the embryonic mushroom. Even at this early stage, the hyphae have arranged themselves into a tiny cap and stem, a homuncular ghost of the mature fruiting body. Following the embryonic phase, hyphae continue to grow, their nuclei multiplying by mitosis and tips extending and branching. But much of the rapid phase of mushroom expansion occurs by inflation of existing hyphae rather than further increase in the number of cells. In comparison with animal tissues, there is little if any differentiation among cell types in most of the mushroom. Indeed, if a piece of stem or cap tissue is placed on agar, the hyphae will grow outward, revert to their invasive behavior, and regenerate a mycelium. Every mushroom cell behaves like an animal stem cell. Something about the position of hyphae within the fruiting body and relative to one another is critical in determining the way that each cell grows, but only while it stays in that place. The only obvious differentiation, other than variations in cell length and diameter in the cap and stem, occurs within the fertile hymenium, where the spores are produced. In the majority of mushroom species, all hyphal apices that emerge at the gill surface become basidia. Basidia must be spaced evenly across the landscape of the gill to ensure that they offer their spores unobstructed launch pads. For this reason, successive generations of basidia arise on the same gills in most species, new ones replacing those that deflate after discharging their quartet of spores. Ink-cap gills are more complex. They house three distinct types of cells: basidia, sausage-shaped paraphyses that establish a sterile pavement between them, and enormous cells called cystidia that inflate across the gap to connect with the opposing gill. The cystidia act as braces, preventing the gills from tearing apart as the cap expands.[8]

We have reached the limits of observational research on mushrooms, and hopes of understanding how fruiting bodies demarcate the cap and stem, establish the spacing of gills before the cap expands, and decide which cells become basidia rest on the analysis of mutants. Mutant strains of the ink-cap *Coprinus cinereus* have been created by disrupting normal genes with short lengths of DNA that are inserted at randomly selected sites in the genome. Fungi that have emerged from this abuse develop bonsai fruiting bodies, form caps or stems that do not expand, or fail to produce spores. One of the great challenges of experimental mycology lies in elucidating the source of the disabilities suffered by these mutants.

The fruiting bodies of many tree pathogens and fungi that degrade fallen logs release spores for months, and there are a few whose hardy brackets persist for years, growing a new layer of hymenial tissues every season. But for other species, the timing of mushroom production is much more exacting. If ripe fruiting bodies of milk caps, or *Lactarius* species, are snapped open, a sap or latex spurts from the cracked cap or stem. Even undisturbed, these fruiting bodies drip latex from the edge of their gills. In the woods close to my home, milk caps emerge in October. After rain I can be certain of finding these mushrooms every year, on the same weekend, at the base of the same trees. Perhaps the most quixotic basidiomycete is St. George's mushroom, *Calocybe gambosa*, which appears in England on April 23 (or thereabouts). Again, emergence is another facet of fungal biology controlled by chemical messages, but this process is complicated by the influence of the environment surrounding the mycelium. Rainfall is an obvious catalyst for fruiting, but the details of other external motivations have not been examined critically.

I have spoken of mushrooms as the tips of mycological icebergs, which may give the impression of a stable, subterranean organism that flushes a little of its biomass to the surface every year. This may be true for the largest mycelia, but for other fungi, the major part of their below-ground cytoplasm is probably shuttled skyward into the fruiting bodies. This makes sense for a mycelium growing within the finite resource of a fallen log or even a single leaf. If the mycelium has consumed the available nutrients, any cytoplasm left in the substrate is doomed to starvation. But wherever a species like my milk cap fruits at the same site every year, some of its mycelium must stay in the ground, lying in wait for a new douse of nutrients. Falling leaves offer an annual pulse of fertilizer for fungi in temperate zones. The development of mycorrhizal fungi is additionally constrained by the nutritional status of their plant associates, adding further dimensions to the suite of signals that control fruiting. Fruiting bodies represent tremendous investments for mycelia, and a plentiful crop may drain the organism's resources. Feast and famine are the governing principles of commercial mushroom cultivation. Initially, a mycelium is feasted on a bed of warm and moist manure, providing ideal conditions for hyphal growth. Then, when the nutrients in the manure are close to exhaustion, the bed is cased in damp soil. Shortly after the addition of this wet blanket, the fruiting bodies expand. Usually, the fungus accumulates enough biomass in the manure to support two or three harvests.

The importance of weight gain followed by starvation as the signal for reproduction in water molds was discovered in the nineteenth century.[9] Wild mushrooms behave in the same way. Although morels are ascomycete fungi, they are large enough to be called mushrooms. Morels are prized edibles, although I believe that many who profess such love for them are greatly prejudiced by the cream sauces in which they are simmered, rather than their scalloped heads and pale gray flesh. They can be found toward the end of April or the first week of May in the eastern United States. Common species include the yellow morel, *Morchella esculenta*, and the snakehead, *Morchella semilibera*. People who would never venture into the woods for any other reason will get up before dawn to collect morels in the pouring rain. These fruiting bodies are not abundant every year, but appear according to long-term cycles that are often misrepresented in local folklore. An elderly woodsman, drawing on eighty years of experience, assured me that every three years brought forth a magnificent harvest in Ohio. I trusted the lines in his face. Four years later, I'm still waiting for morels. What is true of all mushroom forecasts is that the odds of encountering a bumper crop rise with the passage of an increasing number of barren seasons. Mycelia are exhausted by a year of exuberance and then demand time to restore their biomass.

Mushroom hunting has become a serious commercial enterprise in some regions[10] and is a source of recreation for many people, including a large number who call themselves amateur mycologists. Although these hobbyists lack botany degrees, their ability to find and identify mushroom species far exceeds the capacity of someone like myself with years of professional experience in mycology. It is rare in the sciences for amateurs to play such an important role in the continuing vibrancy of a field, and perhaps only astronomers can boast of such a partnership between academia and the enthusiast. But despite the fact that amateur mycologists are responsible for a great deal of our knowledge of mushroom distribution and diversity, I'm not convinced that they should be encouraging the sport of mushroom hunting. Because mushrooms are reproductive structures that decay or are eaten by animals, collectors argue that their activities do no harm. A single mushroom can disperse hundreds of millions of spores in a single day, so picking a few fruiting bodies should have no impact on mushroom emergence in the future. There have been a few careful studies from Europe that support this optimistic view by showing no evidence that mushroom collecting has a destructive effect. But there are many other instances in which a precipitous decline in the annual haul of wild species has been reported. Tuscan truffles are a celebrated cause, and stories of dwindling numbers of certain mushroom species in forests of the Pacific Northwest and other areas cannot be ignored.

The problem, as always, is human population. Although they probably disrupt mycelia by compressing the soil, a few mushroomers combing through a forest cause no lasting damage. But logic also suggests that an ecosystem will be impacted by collectors who strip every fruiting body from the same area every year. In addition to soil compaction, the elimination of spore production by a significant proportion of the fruiting bodies will limit outcrossing through sex and diminish genetic diversity. Just as the Florida panther and other rare mammals may be condemned by the effects of inbreeding, over-picked mycelia may acquire the mycological counterparts of bad teeth, cross eyes, and an unnatural interest in banjo playing. Finally, the elimination of mushrooms as a food source for various animals and as a breeding ground for specific insects may have serious consequences on the viability of forest ecosystems. Habitat destruction and environmental pollution are probably far more catastrophic to fungi than collectors, but mushroom hunting cannot be regarded as a benign activity. I know I'm being selfish; I'd much rather see lots of mushrooms than lots of people. The siren songs from our gonads are driving us toward some very nasty rocks. Global climate change seems an inevitable outcome of our reproductive excess, and will accelerate the departure of the last hominid species from earth. But for mushroomers, there may be a short-lived silver lining to the approaching apocalypse. Freakish weather will lead to some unusual fruitings, and fresh morels at Christmas may ease our well-deserved transit to the thinnest of smears in the geological record.

CHAPTER 8

Angels of Death

Despite popular belief to the contrary, there is only one practical way of distinguishing between edible and poisonous toadstools. This experimental method is sure, but the result may not profit a man.

—John Ramsbottom, *Poisonous Fungi* (1945)

There is a wizened pine tree that curls from the ground beside a sandy path in an area of saw palmetto scrub near Cedar Key, west of Gainesville. I stroll around this rare fragment of Floridian paradise every year if I can, looking for mushrooms, keeping an eye open for the wild hogs that rustle through the dry-leaved thickets, and hoping my prostate will benefit from its proximity to the native vegetation. A mycelium flourishes in the soil beneath the conifer, and a flush of pure white mushrooms surfaces in the fall to shed its spore clouds. The stem of each fruiting body rests in a membranous cup, or volva, and sports a floppy ring just below the cap. These are destroying angels, the ravishingly beautiful mushrooms of *Amanita virosa* (Figure 8.1).[1] The same mycelium has probably fruited in the shade of the old tree for decades, and since the area is protected from development, I'm confident it will continue to do so long after my carcass is ashed. This is a comforting reflection; perhaps I'll lavish my minerals upon the noxious fungus by having my urn emptied beneath the tree. Truly, this angel haunts me.

To a reckless chef, the destroying angel is a formidable siren. Misused as a cooking ingredient, its alabaster flesh has wiped out whole families. The poisonous nature of the mushroom is due to the production of amatoxins, miniproteins or peptides that are absorbed from the intestine and

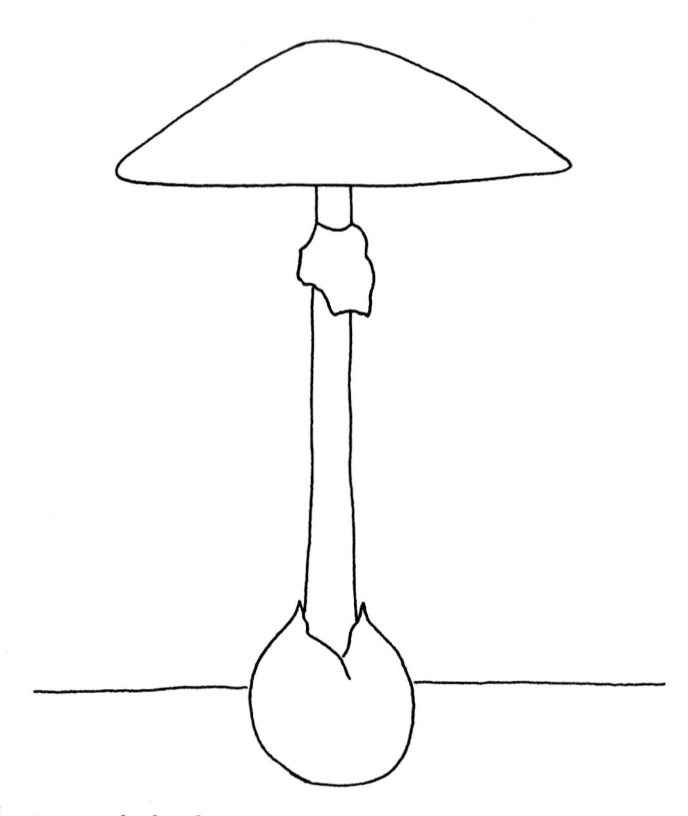

Fig. 8.1 Fruiting body of *Amanita virosa*, the destroying angel. The floppy ring on the stem is the annulus. This is derived from a membrane called the partial veil, which covers the gills until cap expansion. The cup at the base of the stem is the volva, the remnants of the universal veil that wrapped around the whole fruiting body prior to expansion.

lay waste to the liver. In textbook cases of amatoxin poisoning, the unlucky patient can feel fine for some time after consuming the mushrooms. The first symptoms begin eight hours later, or even after a delay of two days, with abdominal cramping, vomiting, and violent diarrhea. But then, the symptoms abate for a while, and the patient may appear well enough to leave the hospital. Unfortunately, this reprieve—called the honeymoon—does not anticipate a rosy future for the patient. Liver damage continues, coma ensues, and death is a frequent outcome.

Dennis Benjamin is a pathologist at the Cook Children's Medical Center in Fort Worth, Texas, and an expert on mushroom poisonings. He estimates that 6 milligrams to 7 milligrams (thousandths of a gram) of pure toxin is sufficient to kill an adult. This corresponds to the dosage delivered by a single *Amanita* cap. On a more positive note, overall mortality rates following amatoxin ingestion are well below 30 percent for patients who receive medical treatment. Patient care centers on the replacement of fluids lost in the early phase of the illness, plus careful monitoring and management of serum chemistry. No antidote for the toxin is available, although French investigators have been claiming miraculous cures since the early 1800s, using potions ranging in refinement from minced rabbit brains and stomachs (sweetened with jam to increase palatability of course—ah, the joys of French cuisine!) to mixtures of antibiotics and vitamin C.

Other species of *Amanita* contain the same amatoxins as the destroying angel. The death cap, *Amanita phalloides*, is responsible for most fatal poisonings throughout Europe, and probably in North America, where the fungus has a relatively limited distribution. Many recent cases in the United States have involved immigrants who have misidentified a lethal mushroom as an edible fruiting body that grew in their original homeland. For example, at first sight, destroying angels and death caps resemble the paddy straw mushroom, *Volvariella volvacea*, which is widely cultivated in Asia. Vietnamese and Laotian immigrants have been felled by this mistake. Recently, a family in Ohio added slices of a pure white mushroom to their soup. The poisonings were so severe that two young sisters had to be treated by liver transplantation.[2] Had the girls not sickened quickly, an epidemic could have occurred: the family owned a restaurant and intended to serve the dish to their customers. Amatoxin fatalities are not restricted to immigrants. Children of all ages are susceptible, and anyone without mycological training that picks wild mushrooms is at risk. The death-cap poisoning in 1997 of Sam Sebastiani, son of the celebrated Californian vintners, was one such tragedy that attracted wide media attention.

The autumn skullcap, *Galerina autumnalis*, is a common wood-rotting fungus that buries its mycelium in fallen logs and resurfaces as a small brown mushroom. The fruiting bodies are easily overlooked: the stems are never taller than a finger or thicker than a pencil, and the caps expand to just a few centimeters in diameter. This is a perfect example of an LBM, or little brown mushroom, one of the multitude of dun, formulaic woodland fruiting bodies that rarely provoke a second glance. But while the guidebooks award some LBMs a cooking pot emblem, a skull and crossbones shares the page with this species. The autumn skullcap contains amatoxins, and its other common name is the deadly galerina.

Occasionally, I am consulted by my local hospital about a patient who has eaten wild mushrooms and has been vomiting. Often, the patient is a child, and the vomiting was induced by an alarmed parent rather than the mushroom itself. Based on the description of the fruiting bodies, I can usually introduce some calm into the emergency room, but given my ineptitude with LBMs I'm fearful of a future consultation when I'll be presented with a handful of something that could be galerina: "He was eating these when we found him." What will I say? "Your son is probably fine, but let's wait a few hours . . . if this is *Galerina autumnalis* he might die." In an effort to avoid this predicament, I decided to spend an afternoon familiarizing myself with the species.

Soon after entering the woods close to my home, a small group of LBMs presented themselves. They looked like the galerina mug shots in my guidebooks, so I plucked them and sat on their log. Flipping through the books I found that edible fruiting bodies of the two-toned scalecap, *Pholiota mutabilis*, are roughly the same size and color, and grow on the same types of decaying wood as deadly galerinas. With growing frustration I flipped through my mushroom guides in search of some distinguishing characteristic that would clinch the identification. Both mushrooms have white rings that circle the stem beneath the cap, but these often disappear with age. The scalecap's stem is clothed with tiny scales, but these are not prominent in older specimens. The stem of this LBM was fairly smooth. Perhaps this wasn't galerina or scalecap. Maybe it was the cluster coincap, *Collybia acervata*, or a species of *Mycena*. One book said that the coincap had a bitter taste, so I nibbled at one of the caps and spit the pulp on the ground. The mushroom was tasteless, although I could smell a very faint cucumber odor as I handled the fruiting bodies. This was a clue. A "farinaceous" smell was mentioned in one of the descriptions of the deadly galerina. The glossary translated this as "cucumbery." Plowing on, through book after book, I became convinced that I was looking at deadly galerinas and had held its tasteless flesh in my mouth. I began to sweat.

Although only a fool would taste-test wild mushrooms without the guidance of an expert,[3] my behavior does not justify immediate institutionalization. Between ten and twenty of the small fruiting bodies would constitute a lethal dose. The smidge of exotic peptides that might have made it into my bloodstream offered little more than mild amusement for my liver.

Poisons produced by other organisms, including marine molluscs, insects, and snakes, are supposed to act as deterrents to predators, and countless plant toxins repel herbivores. But this explanation doesn't make sense when we think about amatoxins. The sting of a bee or a nettle has an immediate effect on animals that might otherwise eat them. Humans and other mammals which eat fruiting bodies are not discouraged by amatoxins (witness recent incidents of mushroom poisoning), because their toxicity is not evident until hours after consumption. A delayed-action toxin is not useful to the organism whose genes have already been digested in an animal's stomach: there is no selective advantage in late retribution. So why do mushrooms produce toxins?

Mushroom toxins are described as secondary metabolites, compounds that are not part of the energy-harvesting and cell-building chemical reactions that are essential to life. It is possible that they are byproducts of more important reactions, unavoidable biochemical trash that performs no useful function. This is the default hypothesis which we adopt in frustration when the synthesis of a metabolite does not appear to confer any obvious benefit. (In Chapter 5, I discussed fungal luminescence, which may be part of the flotsam of cellular biochemistry that is of no particular value for a mushroom.) But other ideas about toxins deserve careful consideration.

An historical perspective—a deep historical perspective—is sometimes brought to bear upon obscure characteristics of organisms that seem to offer no selective advantage. A good example is furnished by the fleshy fruits of the Osage orange, a common tree in North America. These pale green balls are larger than grapefruits and splatter on the ground after ripening. Some authors have claimed that Native Americans had a recipe that made Osage oranges palatable, but nothing touches them now. Even raccoons, opossums, and skunks who show no hesitation in consuming the rotting contents of a trash can shun the Osage orange. The production of a massive and unpalatable fruit is resolved by the idea that the Osage orange was once dispersed by mammoths or other large mammals which have been extinct for millennia. It is a ghost of evolution.[4] However, the notion that amatoxins were designed to dissuade a prehistoric animal with a powerful taste for mushrooms is not compelling. These toxic peptides frustrate protein synthesis by blocking the expression of genes, and I can't think of any animal that would be affected quickly by ingesting something with this mode of action. Furthermore, destroying angels and death caps are often scored deeply by the rasping mouthparts of slugs and snails, and some mammals eat them with impunity.

A better explanation may be found in the effects of the toxins on developing insect larvae that hatch from eggs deposited in the fruiting body tissues.[5] Boletes and some *Russula* species become riddled with insect grubs, and even if the spore-producing parts are not eaten, the active life of the fruiting body can be reduced if the stem or cap is sabotaged (remember that details of the form of the entire fruiting body, including the spacing of the gills, are critical for spore dispersal). High rates of cell division and the attendant processes of tissue differentiation make all embryos very prone to genetic damage, so insect larvae that hatch within the toxin-saturated flesh of an *Amanita* are likely to be highly vulnerable. Poisonous *Amanita* fruiting bodies also contain phallotoxins, compounds that bind to components of the protein skeleton inside cells and inhibit growth. Phallotoxins are not absorbed from the gut, and thus cannot be responsible for any death following *Amanita* consumption. But this second type of harmful peptide may be an additional weapon against insect larvae that are submerged in the toxins. Insects and fungi have interacted with one another for 400 million years or more, an expanse of time that has offered ample opportunity for the evolution of defense and counter-defense mechanisms in both groups of organisms. If insect larvae are the real targets of amatoxins, then human casualties were "unintended" by evolution. In every sense, then, mushrooms couldn't care less about *Homo sapiens*.

Finally, it is worth mentioning the idea that amatoxins were not "designed"[6] as poisons against anyone who might browse on the fungus, but nevertheless fulfill other duties within the developing fruiting body. By limiting protein synthesis in certain tissues, the toxins could affect the manner in which the structure unfolds and begins releasing spores. But just in case this answer begins to seem attractive, it is worth noting that *Amanita* fruiting bodies can be sculpted in the absence of amatoxins. Caesar's mushroom, *Amanita caesaria*, looks very similar to the death cap, but is toxin-free and has been a prized edible for centuries.[7]

Species of *Amanita* produce a remarkable array of compounds that interfere with human physiology. The best known of all mushrooms, the fly agaric, *Amanita muscaria*, is associated with a rich history of ritual use and abuse, and has played center stage in numerous fiction and nonfiction books. The frequently disoriented world described in Lewis Carroll's *Alice in Wonderland* (1865) probably owes much of its detail to contemporary accounts of fly agaric inebriation by the mycologist Mordecai Cubitt Cooke.[8] The visual hallucinations induced by the mushroom are thought to be due to muscimol and ibotenic acid, compounds which, respectively, excite and sedate the nervous system by binding to certain brain receptors. As with the toxins, mushroom hallucinogens were not designed for human use, and we have no idea what service, if any, they perform for the fungus.

In the fall, fruiting bodies of the lawyer's wig, *Coprinus comatus*, emerge from wet lawns as white spindles with rounded tips, extend to 6 inches or more, and unfurl their caps before dissolving ("deliquescing") into black slime (this species was described in Chapter 1). Picked before they begin to blacken, they are quite tasty, but caution is advised. Fruiting bodies of a related species, *Coprinus atramentarius*, look similar but contain a toxin called coprine. If alcohol is consumed when this mushroom is eaten, poisoning is evidenced by tingling in the arms and legs, nausea and vomiting, and palpitations associated with a racing heart. Normally, we convert alcohol to acetaldehyde and metabolize acetaldehyde into acetate. Coprine blocks the second stage of this pathway, causing an abnormal accumulation of acetaldehyde. The symptoms of this metabolic block are identical to those experienced by an individual who quaffs adult beverages while taking a prescription drug called disulfiram (marketed as Antabuse®). For some alcoholics, disulfiram is the only thing sufficiently powerful to subvert their relentless addiction, which provides a hint of the ghastliness of coprine intoxication.

Besides species of *Amanita*, *Galerina*, and *Coprinus*, there are a few other fungi that should never be found in a kitchen. Certain webcaps, species of *Cortinarius*, contain a compound that destroys cells in the kidneys and can induce renal failure. An interesting feature of these poisoning cases is that the delay between ingestion and the onset of symptoms is usually longer than a week, and some patients can appear healthy for up to three weeks. False morels, and a number of related ascomycetes that produce large fruiting bodies, are also to be avoided. Medical histories of false morel poisoning parallel those of patients tortured by amatoxins. Fortunately, it is difficult to mistake a false morel for a real one, and most fatalities occur in people who knew what they were collecting. Apparently, these fruiting bodies can be eaten safely if they are cooked to evaporate the volatile toxin gyromitrin, but I agree with Dennis Benjamin when he questions the sanity of anyone who would play this game of mushroom roulette.

There is a list of other suspect fungi that should be treated with caution, although the effects of these species vary greatly from person to person and at their worst are limited to gastrointestinal distress. In this regard, I will share a humbling personal experience. Students in John Webster's laboratory lived and breathed fungi, sometimes to our detriment, and we ate them too. After a weekend foray, John left the fruiting body of a bolete on my bench. Remembering that he had discussed the culinary delights of mushrooms with me in the previous week, I took the thing home, sliced it into one of my signature student recipes—chili-con-everything—consumed a bowl or two and went to bed. Next morning, John asked me if I had recognized *Boletus satanus*, Satan's bolete. He'd left it for me to identify, not to swallow! Amazed at my evident health, he walked away shaking his head. Satan's bolete cannot kill a mule, but it can make people shit themselves senseless. Either boiling the concoction or some ingredient in my chili had detoxified the mushroom.[9] It is not known which compounds in poisonous boletes are responsible for gastrointestinal irritation, and similar confusion reigns with other fungi. Even the delicious yellow bracket called the sulfur shelf, *Laetiporus sulphureus*, is known to sicken some people. However, before swearing off all mushrooms, it is useful to acknowledge that products as innocent as Wonder® bread probably have some detractors.

Most guidebooks give adequate warnings about truly dangerous species, but they cannot be trusted to deliver a completely reliable recommendation for dinner. Attempting to impress a group of colleagues and students, I once cooked a bracket fungus called the dryad's saddle, *Polyporus squamosus*. Its delightful flavor was mentioned by a couple of authors and following one of their recipes, I made a saffron-flavored dryad's saddle stew. The fresh fruiting bodies emitted a strong perfume, akin to the smell of very cheap cologne. This didn't bode well, but I reasoned that the threads of saffron would dominate the final flavor. But as the broth simmered, the scent from the brackets intensified until it matched the pungency of a disinfectant used in a slaughterhouse. Removing the lid of the casserole when my guests arrived, I forced a mouthful down and attempted a feeble smile. Everyone was horrified. I remain revolted by dryad's saddles and never touch them in the woods. Even the scent they leave on my hands is sufficient to provoke stomach contractions.

Personal appetites vary greatly among mycophagists, of course, but the bizarre tastes of mycologist Captain Charles McIlvaine, who coauthored *One Thousand American Fungi*,[10] are beyond comprehension. He celebrated the delights of earth tongues (tiny black fruiting bodies), phallic mushrooms, earth-stars, and birch conks (brackets with the consistency of wine corks), although even he eschewed false morels. David Arora discusses McIlvaine's psychosis in his book *Mushrooms Demystified*,[11] and also quotes an appraisal of a waxy cap mushroom by an equally eccentric mycologist called Luen Miller:

> Not as good as *Hygrophorus sordidus*, but as an edible species it is not to be despised. It has a noble waxy texture and makes a toothsome meal. Although it lacks the succulence of *Hygrophorus sordidus*, it largely makes up for it in possessing a copious supply of wax, which coats the mouth and throat for hours after eating it, the way good ice cream does.

Fungi assault humans in a variety of ways, either by growing in our tissues with some measure of determination or by pure happenstance on the part of the toxin producers and those with allergic spores. The mold index reported in newspapers, which refers to the concentration of airborne fungal spores, furnishes another example of our invisible but profound intimacy with fungi. While contact with them cannot be avoided without a biohazard suit, some human interactions with their spores are worthy of a "Darwin Award."[12] Take the rare respiratory illness called lycoperdonosis. The name refers to *Lycoperdon*, the puffball genus. In 1994, a group of teenagers from Wisconsin was hospitalized following a party in which they had snorted puffball spores. They were probably hoping for some hallucinogenic experience, and indeed, at least one puffball species does have psychoactive effects if it is eaten. But squeezing a puffball close to your nose is a hazardous adventure. Such vast productions of basidiospores sit inside these dry fruiting bodies, that a spritz up the nostrils delivers millions of the spiny cells into the lungs. Within a few days of exposure the Wisconsin teens developed breathing difficulties, high fevers, and myalgia (muscle pain). Biopsies taken from the lungs of two of the patients revealed acute inflammation of the tissues in contact with masses of basidiospores.

Puffball spores are sufficiently small that they can enter the narrowest airways in the lung and lodge in the alveoli. In this location, macrophages that hunt the glistening lung linings for foreign objects engulf the spores and attempt to digest their prey. But because each spore is equipped with a protective jacket in the form of its chitinous wall, the macrophages suffer indigestion and crawl around with a conspicuous bulge. Studies in which cultured macrophages are allowed to browse on spores show that they remain intact within the macrophages for weeks. But by moving from the alveoli to the parts of the lung that are ciliated, these hunchbacked immune cells sacrifice themselves to the conveyor belt of mucus and are scrubbed from the lungs along with their fungal cargo. Coupled with the use of steroid therapy to control the inflammatory response, this natural defense mechanism allowed all of the teens to recover within a few weeks. The universal teenage pursuit of self-evasion is not responsible for all cases of lycoperdonosis. Intentional limited inhalation of puffball spores is a folk remedy used to stop nosebleeds, and spores were once applied by midwives to the stump of the cut umbilical cord of newborns to stop bleeding.

Ergotism (which was discussed in Chapter 4) is a classic example of poisoning by food contaminated with a microorganism. The dramatic symptoms of this affliction are rare today, but we remain vulnerable to toxins synthesized by other fungi that grow in our food. Aflatoxins are examples of mycotoxins, poisons produced by mycelia (rather than fruiting bodies). Traces of these compounds are present in corn, peanuts, peanut butter, milk, eggs, and meat. Unfortunately, they are among the most potent naturally occurring compounds that induce mutations and cause cancer. They are organic molecules with a ring structure that is perfectly suited for reaction with the DNA double helix, and if they find their way into our nuclei, they bind to DNA molecules (creating an adduct) and cause serious genetic damage. Aflatoxins are synthesized by two species of conidial fungi, *Aspergillus flavus* and *Aspergillus parasiticus*, whose asexual spores are omnipresent and will germinate on plant surfaces. They colonize crops at all stages of development, and also after harvest, but growth is most prolific on plants that have been damaged by insect activity or other types of environmental stress. In the most heavily infested crops each kilogram of plant material can contain as much as 0.1 gram of fungus (which equals one part fungus to 10,000 parts food, and is comparable to the quantity of vitamins in a bowl of breakfast cereal). The harmful resident becomes part of the harvest and its toxins are incorporated into the food chain. Contamination of dairy products and meat is due to the consumption of aflatoxin-tainted food by cattle, and foods of both plant and animal origin contribute to the presence of aflatoxins in human milk.

If rats are fed a dose of aflatoxins equivalent to the maximum exposure anticipated for an American, one of every 10,000 animals develops liver cancer. Circumstantial evidence suggests that the toxin causes the same disease in humans. But surprisingly perhaps, the incidence of the disease in the United States, including cases that have nothing to do with aflatoxins, is tenfold lower than that forecast by the diseased rodents. Clearly, rodents and humans react in different ways to the same toxins, which is little consolation for the caged animal enduring waves of nausea and a painfully distended belly. Incidentally, research on aflatoxins began with an episode of extravagant animal suffering. In England in 1960, 100,000 turkeys died following loss of appetite, lethargy, and liver failure. It was discovered that they had been fed aflatoxin-contaminated peanut meal, and the name turkey X-disease was coined for their plague.[13] The economic loss was considerable of course, and the birds suffered greatly, but the fact that the turkeys faced an unpleasant future with or without aflatoxins may explain why the poultry industry has yet to erect a monument to this terrible event in avian history. I think, however, that a gargantuan stainless steel turkey lying within a circle of faux peanuts would serve as a very tasteful memorial. Tourists would enter the statue through the open beak, and exit through its rear end with new respect for the fungal enemy that pursues us all.

As I've indicated for other fungal toxins, it is unlikely that turkeys, or any other animals, are the targets for aflatoxins in an evolutionary sense. It has been suggested that animals compete with fungi for the same food—harvested grain, for example—and that aflatoxins would reduce the number of these rivals. But I'm not convinced; other microorganisms are probably the "intended" victims. While fungi produce mycotoxins in minute quantities in the soil and in plant tissues, local concentrations may be high enough in the immediate vicinity of the growing hyphae to clear the territory for the mold. The same natural role has been suggested for antibiotics such as penicillin and cephalosporin, which we employ against bacterial infections.

By secreting antibiotics, some fungi benefit not only by removing the competition, but also obtain an extra shot of nutrients in the form of the dead bacteria. Fungi are confronted with a serious nitrogen shortage when they consume plant tissues because these food sources yield plenty of sugar but negligible protein. Without proteins or an alternative source of nitrogen-containing molecules, fungi cannot build their own proteins or nucleic acids (DNA and RNA). To augment their vegetarian diets, mushroom-forming basidiomycetes attack soil bacteria with secreted antibiotics and absorb nutrients from their leaking cells. Interestingly, the mycelia of the species that form ink-caps, puffballs, and bird's nests are all adept predators of bacteria. Other kinds of fungi obtain nitrogen by trapping and killing nematode worms. A single gram of soil can house as many as 1,900 microscopic nematodes and more than 1,000 meters of hyphae.[14] At this density there is no shortage of encounters between hyphae and worms. The simplest system involves short hyphal branches that exude a powerful adhesive. Passing worms become cemented to the tips of these snares and in their attempts to escape become stuck on adjacent branches. Sometimes, the branches detach from the mycelium, but the animal's fate remains the same. Within hours the fungus penetrates its skin (called the cuticle) and dissolves the internal tissues. Other traps are more elaborate, ranging from adhesive cages to explosive rings whose interior wall inflates when disturbed by a worm. Nematodes are lured to these ring traps by chemical attractants—siren songs—and as the animal glides through an opening, the trap inflates and grips the animal with a pressure of up to 18 atmospheres.

While everyone is familiar with poisonous mushrooms, and aflatoxins may represent an underappreciated menace to human health, nothing fungal has captured people's imagination recently like the black mold *Stachybotrys chartarum* (also known as *Stachybotrys atra*; Figure 8.2).[15] This conidial fungus is proclaimed in the tabloids, and in some more reputable newspapers, as a toxin-producing killer. Its current reputation is founded on its activities in 1993 and 1994 in Cleveland, Ohio, where its toxic metabolites, called macrocyclic trichothecenes, caused blood vessels to hemorrhage in the lungs of ten infants, one of whom died. The evidence of fungal responsibility for this incident is statistical in nature. A comparison between children that suffered pulmonary hemorrhage, with a group of children of similar age that were unaffected, showed that the symptoms were associated with living in water-damaged homes infested with *Stachybotrys*. Subsequently, the county coroner reexamined all infant deaths in Cleveland in the early 1990s, and found that another six children may have died of the same cause. Originally, these fatalities had been classified as instances of sudden infant death syndrome, or SIDS. Understandably, there is a great deal of interest in determining the prevalence of the problem elsewhere.

Susan Kaminskyj is a biologist at the University of Saskatoon and is one of the few professional mycologists in the entire Province of Saskatchewan. She receives frequent phone calls from distraught Canadians, terrified by the discovery of streaks of black mold on their wallpaper. Although the problem can be serious, she soothes her callers with the following phrase: "If this mold didn't kill you yesterday, it's unlikely that you'll die today." There is undeniable logic to Susan's advice, but she doesn't suggest that the homeowners ignore their mold. Isolated patches can be treated with diluted bleach, but unless the moisture that created the problem is subdued the fungus will be back. Look at the wall surrounding your bathtub or shower.

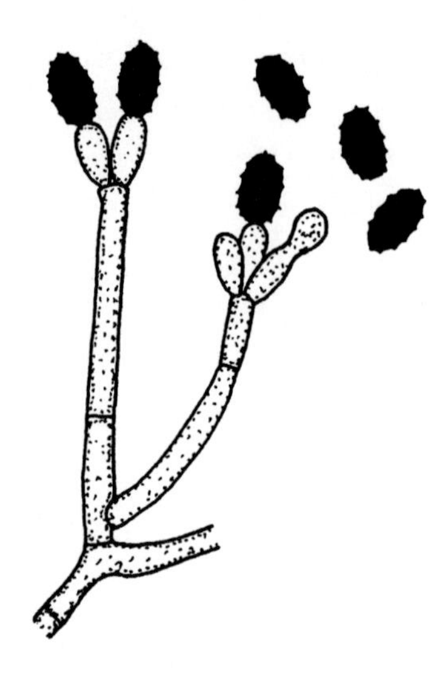

Fig. 8.2 Conidia and conidium-producing stalks (or conidiophores) of the black mold *Stachybotrys chartarum*.

Chances are good that you'll see black stain, the melanin pigment that betrays the presence of these fungi. Usually, this kind of limited mold growth presents no threat to one's health. The horror begins when a basement is flooded, if a burst water pipe remains undetected, or most commonly when drywall is converted to "wetwall" and remains sodden.

Part of the trouble lies in the fact that we live in such well-insulated homes. Drafty as they are, older houses can be very effective at protecting themselves and their occupants from molds. The modern polymer-wrapped house with hermetically sealed windows and other energy-saving features creates a living space with very stale air. The space between different layers of wall materials can become extremely humid when there is no opportunity for water vapor to escape. Air conditioning exacerbates the moisture problem, and may help explain the prevalence of mold damage in certain regions (such as the Houston area), where chilled air is circulated all year round. Having said this, the panic associated with the appearance of black mold is often exaggerated. *Stachybotrys* is not the only black mold that grows in homes. Species of *Cladosporium*, *Penicillium*, *Aspergillus*, and *Alternaria* with melanin-impregnated cell walls are far more prevalent in houses damaged by water, although none of these fungi generate the toxins. Even if *Stachybotrys chartarum* is identified, this does not signify imminent doom, since only a few of the varied strains of the species produce trichothecenes.

Aside from trichothecene production, the spores of indoor fungi present a serious threat to asthmatics. During summer months when fungi are flourishing outdoors, 500 or more spores may drift around in a cubic meter of indoor air (about the volume of two bathtubs). Current standards for air quality suggest that the presence of more than 50 spores per cubic meter presents a significant risk for eye, nose, and lung irritation (although the proposed limit for office buildings is 200 spores per cubic meter).[16] In mold-damaged homes, spores can number in the thousands per cubic meter! The issue is complicated further by the weaknesses in the sampling methods used to determine the level of fungal particulates in the air. In addition to counting spores under a microscope, fungi are detected by opening culture plates for a specified time interval and counting the colonies that develop after a few days of incubation. Because only intact spores will germinate on a culture plate, this method can underrate the total number of irritating fungal particles by a factor of 100. This presents a dilemma because dead fragments of broken hyphae can be just as allergenic as complete spores. Too many variables affect the indoor environment to provide a meaningful average for the density of fungal particles in an American home, but given their ubiquity, there is no doubt that we inhale and exhale fungi and their cellular debris from our first breath to our last.

The majority of the fungal toxins are not serviceable as biological weapons because their effects on the human body develop too slowly. It is doubtful that military strategists have ever considered a warhead loaded with poisonous mushrooms, although there is evidence that Iraqi scientists experimented with concentrated aflatoxins before the Gulf War. The damaging effects of aflatoxin ingestion do not seem to be apparent for many months or years, so I'm also confident that the Iraqi military recognized the limited utility of contaminated peanuts as a battlefield weapon (unless they were planning to strike turkey farms). But what if aflatoxins were disseminated as an aerosol so that they would be inhaled? They might have completely different effects on the lungs.

A clear threat is posed by black mold trichothecenes because they can act immediately upon contact. While there is a great deal of disagreement about whether fungal toxins have been used in various conflicts, there is no doubt that trichothecenes would work if dispersed in significant quantities. The obvious method of delivery would involve an aircraft (even something as primitive as a crop-duster could be frighteningly effective), carrying tanks filled with purified toxins. Trichothecenes cause burning and blistering of the skin, and damage can be sufficiently extensive that whole patches of tissue are shed before the agonized victim dies. Soap and water will remove mycotoxins from the skin, but only if they are used immediately after exposure (trichothecenes are among the biological and chemical warfare agents that can be treated using a special decontamination kit issued to all U.S. military personnel). Inhalation of the trichothecenes results in a severe form of the same symptoms associated with black mold spores, including coughing, breathing difficulties, and a burning sensation in the lungs. Contact with eyes causes tearing and blurring of the vision, and ingestion will result in nausea and repeated vomiting, followed by bloody diarrhea. A few milligrams of the most potent trichothecene mycotoxins, including T-2 toxin and diacetoxyscirpenol, are lethal. The effects of lower doses are not immediately obvious, but a few days after exposure, patients become anemic, platelets become scarce, and plunging white blood cell counts indicate the crash of their immune defenses. There are no specific antidotes for trichothecene poisoning.

Reputable investigators are convinced that these compounds were used by the Soviet Union to kill thousands of anticommunist guerrillas in Laos and Cambodia in the 1970s, and against members of the opposition to their invasion of Afghanistan in the 1980s. A yellow crystalline deposit appears when trichothecenes are purified from fungal cultures. This may account for the color of the putative weapon used in Southeast Asia which was dubbed "yellow rain." Despite the compelling testimony of some victims, a number of recent investigations have concluded that droplets of yellow rain did not originate from an enemy aircraft, but were deposited by swarms of defecating bees. This seems too ludicrous not to be true. Any agencies intent on sidestepping the truth would surely have thought of something more sensible. The grounds for favoring trichothecene delivery in Afghanistan during the 1980s are similarly murky, although more than 3,000 deaths in the region have been linked to the use of an unidentified chemical or biological weapon.

Trichothecenes have also been considered as a possible cause of the some of the devastating illnesses contracted by American servicemen who served in the Persian Gulf in 1991. Declassified documents from the U.S. Department of Defense report that scientists in Iraq were working with mycotoxins in the 1980s, leaving little doubt that they had the capacity to stockpile these compounds for use in biological weapons. It is a depressing fact that the techniques for producing and storing mycotoxins are well within the grasp of anyone with a Ph.D. in microbiology or mycology, provided that the appropriate strains have been acquired and a modern lab is at their disposal. The production of mycotoxins is an incomparably simpler project than any effort to develop nuclear munitions. Western intelligence agencies are very familiar with some of the specialists in biological warfare working in the Middle East, because these scientists earned their degrees from universities in Europe and the United States.

I began this book as a love story, hoping to convince every reader of the awesome beauty of fungi. It is a point of great irony to me that some of the objects of my desire—my Cedar Key angels, for example—are also instruments of a particularly unappealing death. Collectors of firearms might recognize the same contradiction when stroking the latest titanium-clad Italian pistol. Can the toxicity of a few species explain why so many people express revulsion at the mere mention of a fungus?

The other day, a student told me that she had found a patch of giant puffballs on a trail that snakes around my university campus. Although giant puffballs are fairly common, I never tire of seeing them for myself. Birders evince the same sort of eagerness when they look for swallows each spring. I found the fruiting bodies after a short walk. Six snowy orbs, ranging in size from softballs to basketballs, were nestled under some scraggly bushes. They are such endlessly weird things to look at. The largest of the colony had tempted someone to push through the prickly vegetation and kick it to pieces. If the fruiting body had been mature, the vandal would have obliged the puffball's mechanism of dispersal and would probably been very impressed by the resulting spore cloud. But these were young fruiting bodies, a few days shy of transforming their ivory meat into billions of brown spores. Giant puffballs are especially alluring targets, but why is it that so many of us harbor an impulse to kick mushrooms of all kinds? Perhaps some mistrust is buried in the subconscious, an instinct that compels us to act with aggression when we come across a fruiting body, but sparks no similar impulse for molestation when we walk around a tree sapling. Could the fact that fruiting bodies have picked off unwary apes on our bloodline for so long have wired humans against mushrooms, in the same way that most of us shudder when we encounter large spiders? This is a remote possibility. A simpler and more plausible reason is that mushrooms are cursed by the tactile pleasure obtained from snapping their stalks, fragmenting their caps, or from rupturing the skin of a puffball. Kick a sapling and it springs back; boot a mushroom and enjoy the briefest exhilaration, the humblest of highs, as the water-saturated fungus explodes into the air.

CHAPTER 9

Mr. Bloomfield's Orchard

And this stillness of life did not in the least resemble a peace.
—Joseph Conrad, *Heart of Darkness* (1902)

Fungi invaded my childhood by consuming an untended apple orchard on the property next to my home. The orchard belonged to Mr. Bloomfield, the village milkman who lived with his ancient mother in a crumbling house that shadowed a brick-walled dairy stacked with crates. The gnarled branches of Bloomfield's century-old apple trees formed a solid canopy, beneath which sickening fruits swelled and decayed. Rodents stuffed themselves with the fermenting apples, wasps flourished, and birds hopped around the tree limbs taunting a pride of shabby cats. The orchard's lush ceiling trapped a perpetual fog, and in this water-saturated atmosphere trees and fruit were devoured by bitter rot, black rot, blossom end rot, canker, rust, powdery mildew, rubbery wood, and scab. Mushrooms of all colors sprouted under the diseased branches and on the tiny patch of boggy lawn surrounding the house: a pink-gilled *Agaricus*, scarlet waxy caps, and masses of ink-caps that bled into the grass.

Just over our garden fence, this mycological wonderland extended into impenetrable darkness, and mushrooms inhabited my dreams before I knew what they were. My sister and I ventured into Bloomfield's jungle every now and then, always with hearts beating fast at the thought of the milkman and his pair of black hounds. The most magical inhabitant of the orchard was an animal whose drumming I had heard since birth. I saw it one evening, never again. Clinging to a tree at the orchard's edge was a green woodpecker, a magnificent dinosaur that gorged itself on the insects that bored beneath the bark of the dying trees. The combined onslaught from fungi, insects, and birds was dissolving an Eden created by some long dead Victorian, but Mr. Bloomfield didn't seem to care.

Apple trees and all other plants are attacked by bacteria and viruses, nematode worms, and insects, but fungi cause more plant diseases than all other enemies combined. Since the beginning of agriculture, billions of livelihoods and lives have been lost to rusts, smuts, bunts, mildews, potato blight, and rice blast. Of particular significance to me is a pestilence called coffee rust that ravages an indispensable crop in South America. Annual losses to agriculture caused by fungi are incalculably high, and chemical fungicides represent a tremendous investment made by farmers and paid for by consumers. Fungi also spoil food after it has been harvested, rotting fruits and vegetables during storage, transport, and in the recesses of our refrigerators.

Fungi are at war with the plants in my Ohio garden. An ornamental hawthorn is attacked by a rust in wet years. The leaves become covered with yellow spots and some of the red berries on every twig are transformed into pale galls covered with spore-filled horns. The rust produces masses of spores at the base of each of these hollow chimneys, so that elongating towers of infectious cells are driven toward their openings, continually replacing the uppermost spores that escape into the air. Hawthorn rust is caused by a fungus called *Gymnosporangium globosum* that also infects fruit trees and evergreens, including cedars and junipers. The nearest apples are less than 100 feet away, a pair of lively youngsters in my front yard, and there are plenty of ornamental junipers. By showering the spring and summer breezes with spores, the rust has no problem finding food. Although its diet of diverse tree species argues for a broad array of infection mechanisms and digestive enzymes, *Gymnosporangium* (like most plant pathogens) cannot compete with the dietary flexibility of *Cryptococcus* that dines on bird droppings and human nervous systems.

Gymnosporangium behaves in a similar fashion on hawthorns and fruit trees, but produces entirely different kinds of spores when it grows on evergreens. Its life on hawthorns and its life on junipers are separate phases of the rust's existence, which we can summarize in a life cycle diagram (Figure 9.1). Although this provides a succinct summary of a whole organism, generations of biology students have dreaded these

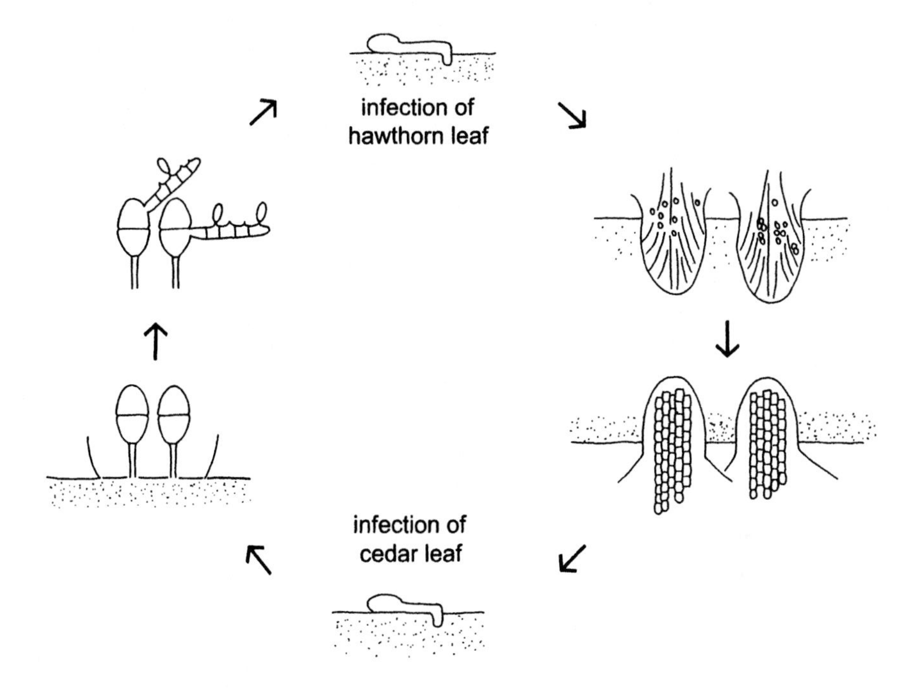

Fig. 9.1 Life cycle of *Gymnosporangium globosum*, hawthorn rust (or cedar-hawthorn rust). The different stages in the life cycle are similar to those described in the text for black stem rust of wheat and shown in Figure 9.2.

creations. The source of the confusion lies in the distinction between the individual organism and an impersonal, bigger picture. The circle seems to imply the endless and pointless conversion of a single organism from one form to the next and back to the beginning, when in fact we are all aware that there is a beginning and an end to every life. When an organism reproduces sexually, part of its genetic heritage survives, but no individual survives beyond a single rotation through the cycle. A spiral may be a more useful diagram than the usual circle.

A familiar example may be helpful. Men and women are built from cells that contain two sets of chromosomes, one inherited from each parent. These cells are referred to as diploid. In the testes and ovaries, sperm and eggs are generated through a special process of cell division called meiosis. Meiosis halves the number of chromosomes in each of the resulting cells and shuffles the genes around into novel combinations. Then, sperm and egg unite to create a zygote in which the diploid number of chromosomes is restored. A great deal of cell division and differentiation ensue, years pass, sexual maturity is reached, and then the reduction of chromosome number can be repeated with the formation of eggs or sperm. At this moment, every one of us fits somewhere on this diagram, and this diagram represents a single turn of the human life cycle. Humans reproduce according to a very simple pattern in comparison with the alternation of sexual and asexual phases in ascomycete fungi discussed in Chapter 4, and the life cycles of some rusts will blow your mind.

Puccinia graminis causes black stem rust of wheat and other cereal crops. It also causes barberry rust, and provokes headaches among botany students because it has multiple distinct stages in its life cycle (Figure 9.2). Only one of these, which develops on the wheat plant, is actually rust-colored. This is the uredial pustule, an open sore that bursts through the surface of the leaves and stem and exposes reddish, spiny spores called uredospores. In a heavily diseased crop, a cloud of uredospores blankets the field after a wind gust, infecting every plant in the field. With permissive winds, spores can be carried hundreds of miles, so that one infected crop can spawn an epidemic.

To infect a healthy plant, the uredospore germinates on its waxy surface, sending out a tiny hypha or germ tube. On the wheat leaf, the germ tube hunts for an opening and is programmed to know where to find one. The leaf has two sets of ridges along its length. You can feel the tall ones by drawing the leaf blade between finger and thumb. These mark underlying veins that carry water and food through the leaf. But the leaf surface is also rippled more delicately, its epidermal cells creating undulating files of microscopic hills and valleys. At intervals along these files, pores called stomata open and close to regulate gas transfer and water loss. When they are open, the plant is vulnerable because each stoma (or stomate) is a portal to the succulent tissues within. Rust uredospores germinate in the morning when humidity is highest and the stomata are open (the plant closes them later as the air temperature rises). Recognizing a valley formed by the junction between epidermal cells, the germ tubes grow across the leaf, rising over the parallel hills and falling into the intervening valleys. If the germlings took the alternative route, growing along the first valley they encountered, they might never find an opening, because one in five, or even fewer lines, of epidermal cells contain stomata. By

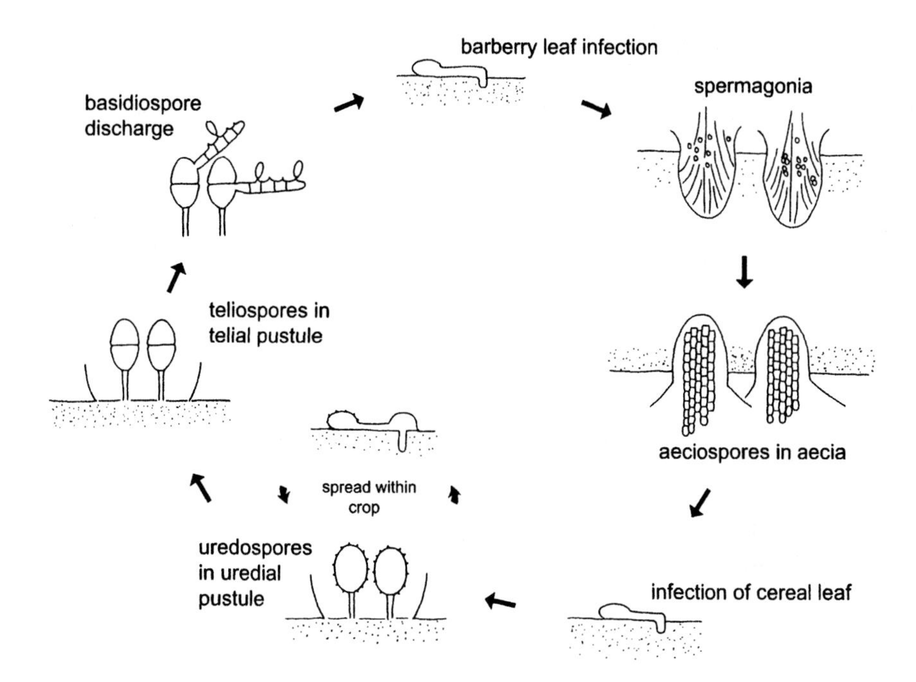

Fig. 9.2 Life cycle of *Puccinia graminis*, the cause of black stem rust of wheat.

maneuvering itself across the leaf, the rust is far more likely to find a stoma (Figure 9.3).

When the tip of the germ tube detects the thin lips around a stoma it inflates over the opening and plunges into the moist interior of the leaf. But how does the fungus know that it has found an epidermal hill or valley, or the lips of a stoma? The lips are raised above the leaf surface by less than a millionth of a meter; a stray bacterium is taller. In a series of beautiful experiments, Cornell plant pathologist Harvey Hoch made plastic replicas of leaf surfaces and found that germ tubes recognized and swelled over his model stomata. This was significant because it demonstrated that the fungus could find its entry point using physical cues alone and did not need to detect the flux of gases through the stomata or perceive some other signal that it was close to an opening. Harvey went further, fabricating films with microscopic ridges of defined height. On these counterfeit landscapes, the fungus recognized ridges that matched the height of stomatal lips (0.5 μm) but crawled over lower or higher ones.[1] Once

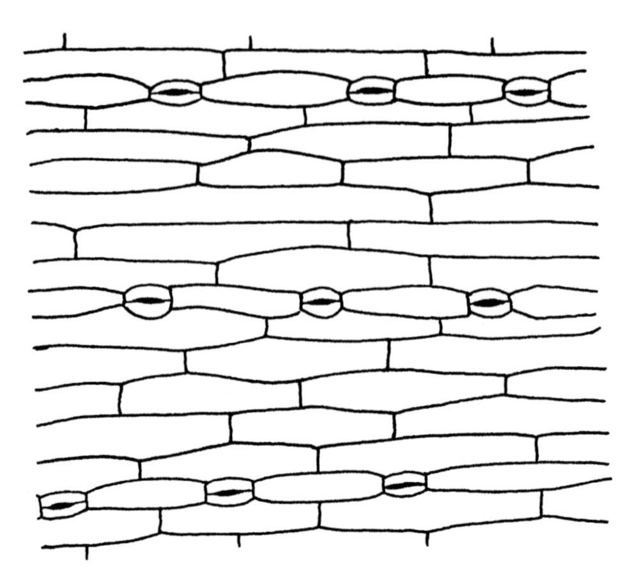

Fig. 9.3 Diagram showing distribution of stomata (pairs of tiny cells that form lips) on the surface of a cereal leaf. By growing across the leaf—top to bottom in this diagram—a germling is more likely to encounter a stomatal opening than one that explores the length of the leaf. In this diagram, only one of every six or seven parallel files of leaf cells contains stomata.

he understood the physical features that the rusts used to negotiate a leaf surface, Harvey manufactured mazes for his rusts, circuits for which he could predict a growth pattern before he seeded the surface with spores. Brilliant research! Robby Roberson, at Arizona State University, used the same methods to control the growth pattern of the fungus, and once a mycelium developed he passed electrical currents through its hyphae. Robby dreamed of producing a biochip—a living, breathing fungal computer. As you know, airlines now use Robby's mycocomputers to schedule all flights. (Could this explain why I have to fly south to Memphis from Cincinnati when I want to visit Toronto?)

The cellular mechanisms that allow rusts to perceive minuscule features of the topography of a leaf surface are not well understood, but may involve proteins that form channels through the fungal membrane. These channels appear to open and close in response to stretching, as the hypha forces itself up and over a ridge. Channel opening allows calcium to flood into the cell, setting off a cascade of biochemical reactions that "inform" the fungus that it has encountered an obstacle of a particular size. These processes are related to the mechanisms that provide us with our tactile sense. Channels in the nerve cells in our finger tips open and close as we run our fingers over a grass blade, causing nerve impulses that signal contact with those varicose leaf veins. But we are not programmed to feel epidermal hills or stomatal lips; there would be no advantage to such finely tuned powers of perception for a human. On a molecular scale, our contact with the leaf is less intimate than that made by a tightly appressed hypha. Our nerve endings are spaced much more widely than stomata, and there is too much noise from our fingerprints. The tactile sensitivity of humans and fungi is tuned to our distinctive needs.

Successive rounds of mycelial proliferation within the plant followed by pustule formation and uredospore production signify a bleak future for a crop. This stage of the life cycle can be likened to the activity of a photocopier, because the fungus is cloning itself in the form of immense numbers of uredospores. It is a very effective method for destroying the monocultures of genetically identical crops that are the lifeblood of modern agriculture. Rust species often encompass a number of special races, called formae speciales (f. sp.), which target particular crops. *Puccinia graminis* includes f. sp. *tritici* that infects wheat, f. sp. *avenae* that attacks oats, and f. sp. *secalis* that is the source of nightmares for hypochondriacal rye plants. But there is further specialization among pathogen and prey because not all varieties of a crop species are equally susceptible to infection. If a rust lands on a resistant plant, a hypersensitive reaction takes place in which the plant destroys its own cells around the point of penetration, forming a tiny fleck in the epidermis. This deliberate cell death strategy is programmed into the plant and is a crucial defense because it starves the fungus of living cytoplasm and prevents spread of the infection. Evolution favors the emergence of new races of rusts that can overcome the defense mechanisms of the plant, and in turn, the plant alters its response to the fungus through natural selection or via artificial selection in the greenhouse. This mutual evolution of pathogen and prey is aptly described as an arms race, just like the war between insects and their pathogens described elsewhere in this book.

Plant breeders select for resistant crops, and their efforts can be informed by the kind of basic research carried out by Harvey Hoch. Once we appreciate that the rust recognizes the microscopic topography of a leaf surface to enter its prey, leaf roughness becomes a potentially important character for selection in the greenhouse. But rather than embarking on years, even decades, of selective breeding, many plant biotechnologists are enchanted by the prospect of direct genetic engineering of crop plants with built-in antifungal defenses. They dream of mutant plants whose epidermal cells produce fungicides. The beauty of these techniques is that they might allow us to reduce fungicide applications from tractors or crop-dusting aircraft, but a vital debate about genetically modified crops must continue for many years. I would be more enthusiastic about plant biotechnology if I believed for one second that the farming of genetically modified (GM) crops would save the life of a single starving child, or even if these plants would reduce the ridiculous cost of breakfast cereal in the United States.

Black stem rust refers to black streaks or open sores that develop on the infected cereal plant and contain a second spore type called the teliospore. (If you had put the book down to make some tea, I should emphasize that I'm still discussing the life cycle of *Puccinia graminis*. This is the transcript of one of the trickiest verbal operations I perform upon students.) Teliospore factories develop within the reddish uredial lesions toward the end of the growing season when uredospores have done their worst. In *Puccinia graminis*, each teliospore is a pair of swollen cells attached to a stalk; imagine a lollipop with two circles of candy, one on top of the other. The spores have thick pigmented walls and are not designed for dispersal by wind. Instead, they remain in their sores and survive the winter months in the frozen stubble. The warmth of spring triggers the next phase of development. Both cells of the teliospore can germinate to produce short hyphae that stick into the air. A nucleus in each hypha is divided by meiosis into four daughter nuclei, and these are packaged into bean-shaped spores that sit at the tips of spikes. Then comes a surprise, one of those piercing insights made possible by Buller's lonely hours at the microscope: a droplet of fluid appears at the base of every spore before it is launched from its spike.[2] The appearance of Buller's drop immediately exposes the bloodline of the rusts. This third type of rust spore must be a basidiospore, which establishes the fact that rusts are basidiomycete fungi, distant cousins of mushrooms. Notice that the nucleus within each rust basidiospore is produced by meiosis, the same type of nuclear division that makes the nuclei of eggs and sperm in animals. If two basidiospores fused, this would be a fertilization event, but no such simplicity is in store. Over the past 100 million years or so since the evolution of grasses, the rusts have fashioned a Fabergé egg of a life cycle.

The basidiospores are dispersed in air currents swirling above the stubble, and survival of the fungus is dependent on delivery to a different plant species, a thorny bush called the barberry. Some infinitesimal proportion of basidiospores hit a wet barberry leaf, germinate, and penetrate the surface immediately, pushing through the cuticle and underlying epidermal cell wall without searching for an open stomata. The cell walls of grasses like wheat are toughened by the presence of silica, creating a hard, glassy barrier to invasion. Presumably, this was an important factor in driving the evolution of the surface-sensing mechanism in rusts that allow them to locate stomatal openings. Barberry leaves are not as tough as those of cereal crops, making it easier for the fungus to puncture this plant by secreting wall-degrading enzymes and applying mechanical force.

Once inside the barberry, a mycelium grows between the cells of the leaf. The fungus feeds by pushing its way into these cells with bulbous branches called haustoria. Haustoria breach the walls of the leaf cells but do not break the plant's membranes. Instead, the membrane of each infected cell is dimpled so that it fits like a glove around the haustorium, creating a placenta-like connection between pathogen and host. The placenta analogy is useful, because in both mammals and rusts, a tight physical connection is created between two different organisms, and second, because this allows the parasitic partner (baby) to absorb nutrients from the host (mother). Rusts also generate haustoria when the mycelium is established in the cereal host. This feature of the infection keeps the host cells alive, prolonging the parasite's banquet. Other fungi lack this finesse and obliterate plant cells with waves of enzymes and feed on the resulting pulp.

The mycelium that develops from a rust basidiospore contains identical nuclei derived by mitotic division of the spore nucleus. Rust sex allows one of these mycelia to fertilize another, and this fertilization occurs on the barberry bush. Perhaps another cup of tea is in order, but before you hoof off toward the kitchen I will enlighten you with the fact that rusts cannot make love without the active participation of insects.

When the rust has grown within the barberry leaf for a few days, it forms tiny flasks called spermagonia that rupture the upper epidermis. Thin hairs protrude through the opening and each spermagonium exudes a sugary nectar and masses of spherical cells called spermatia. Spermatia should not be called spores because they cannot germinate and produce a mycelium. Instead, they function as sperm cells and the spermagonia can be likened to flowers. In their frenzied search for rust nectar, flies and other insects carry spermatia from one spermagonium to another (Figure 9.4). Insects feeding on the nectar brush spermatia onto those hairs that poke from the spermagonia, and in a successful cross, a spermatial nucleus is injected into the hair of a mate. Transfer of the nucleus is followed by its migration down the hair and repeated division of the nucleus. Deep in the barberry leaf, a mycelium that contained identical nuclei is transformed into a mycelium with two different types of nuclei. This structure is called a dikaryon, just like the mycelium of other basidiomycetes which is established before the production of mushrooms. The flower analogy is strengthened by the discovery that some rust infections alter the appearance of the host plant so that the moldy stalks resemble normal flowers. In the case of the rust *Puccinia monoica*, that infects plants in the cabbage family, clusters of leaves covered with yellow spermagonia are such perfect mimics of the actual flowers that fewer of the surrounding plants of other species with yellow-colored flowers become pollinated and set seed.[3] The fungus satisfies its sexual needs by exploiting the insects' insatiable appetite for yellow advertisements, and in so doing, changes the ecology of its entire habitat.

A single spore type completes the life cycle of the fungus that causes black stem rust. On the underside of the barberry leaf, the dikaryotic mycelium forms cup-shaped cavities called aecia that sprout columns of aeciospores. When an aeciospore lands on a wheat leaf, it germinates and penetrates the plant through an open stoma, if it finds one, or sometimes does so directly (somehow overcoming that glassy barrier). Then it feeds with haustoria, and spreads between the cells for a few days before uredial pustules burst through the leaf. Another crop infection is under way.

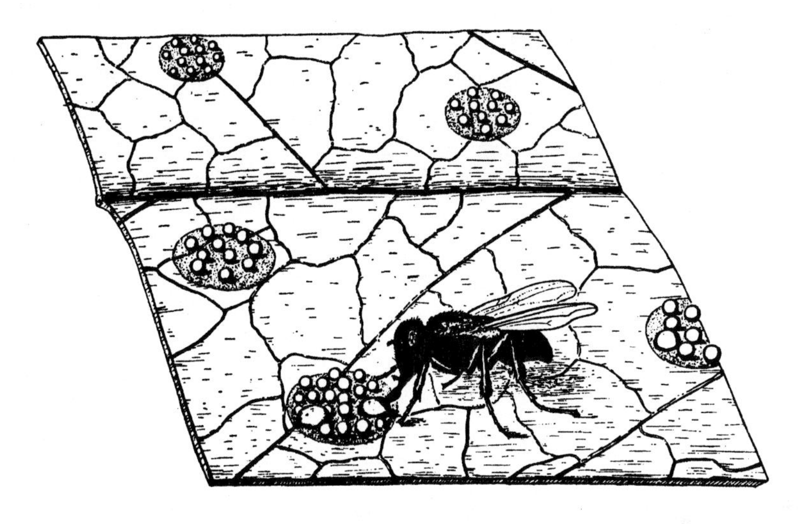

Fig. 9.4 Fly drinking nectar from rust spermagonium on barberry leaf. Reproduced from A.H.R. Buller, *Researches on Fungi,* vol. 7 (Toronto: Toronto University Press, 1950).

Two hosts, four types of spore, the counterpart of sperm cells, and three acts of plant penetration. Why is this life cycle so complicated? The same question can be asked of parasitic worms that require two animal hosts, or of the malarial parasite that multiplies in the salivary glands of mosquitos and also bursts human red blood cells. The standard answer centers on resilience. The opportunity for survival is boosted by the availability of two hosts with different growing seasons. Wheat and barberry are not closely related, but the alternate hosts of some rusts are even more estranged. One rust infects fir trees and ferns, another attacks white pines and gooseberries. Anyone with botanical training will appreciate this omnivory. Multiplication and, secondarily, recombination of genes through sex are keys to survival, and the exploitation of two hosts may maximize the opportunities for both biological activities. Another remote possibility, founded on pure speculation, is that rust species are hybrids formed from two distinct parasites that once lived on different hosts. Similar ideas have been put forward to explain the byzantine life cycles of marine invertebrates like echinoderms, and recent research on lateral gene transfer makes this dream slightly more plausible.

Rusts can be controlled with chemical sprays that range in sophistication from sulfur-containing mixtures to synthetic fungicides that permeate the whole plant and target specific metabolic processes. Some of the synthetics share a similar mode of action with the antifungal agents used to treat human mycoses. For example, black stem rust can be controlled with triadimefon, a compound that disrupts membrane function by inhibiting ergosterol biosynthesis. Cultivation of rust-resistant wheat cultivars is a more effective approach that reduces the cost and environmental pollution associated with fungicide application. Prospects for genetically engineered varieties of crops with customized biochemical pathways for antifungal defense are both exciting and terrifying. Losses from black stem rust have also been reduced by eradicating barberry bushes from wheat-growing regions. Since sexual reproduction occurs on the barberry leaf, the absence of this alternate host retards the evolution of new strains or races of the fungus. This, in turn, lengthens the career of a particular wheat cultivar.

Many rusts have edited versions of the life cycle described for *Puccinia graminis*. Coffee rust is caused by *Hemileia vastatrix*, a fungus that produces uredospores, teliospores, and basidiospores, but lacks spermatium- and aeciospore-producing stages. Remember that in black stem rust, spermatia and aeciospores are produced exclusively on barberry, the alternate host. The edited life cycle of coffee rust is due to the fact that the fungus thrives on coffee alone. It has no alterative food source. This deviance makes sense to me every morning, pre-espresso, as I stand in front of the bathroom mirror, my hair styled, apparently, by a hand-grenade. Although I suppose that life can be maintained (however miserably) in the absence of coffee, the importance of coffee growing to local and national economies in South and Central America, Africa, and Asia places considerable urgency upon the introduction of effective control measures. Without the luxury of an alternate host like barberry, whose removal might control the disease, prevention of coffee rust is limited to fungicide spraying and the cultivation of resistant crop varieties.

More wheat is cultivated than any other crop, making black stem rust a celebrity among pathogenic fungi. Since rice ranks as the second largest crop on Earth and is a staple food for more than half the world's population, its enemies also deserve serious investigation. A rice field infected by *Magnaporthe* looks as if it has been struck by a bomb, which accounts for the common name for the disease: rice blast. The fungus usually infects the leaves of the rice plant and spreads as an invasive mycelium, consuming the plant's tissues and starving the developing grain. Sometimes, the neck or panicle of the rice plant is ravaged by the fungus, becomes bleached by the sun, and withers to a dry straw. Because the grains normally ripen on the panicle, crop loss due to "neck blast" is severe. Blast is most prevalent in areas of intensive agriculture where the soil is supplemented with nitrogen fertilizers; crops grown by subsistence farmers who rely upon natural soil fertility are less likely to be devastated by blast. The only positive thing I can say about rice blast disease is that scientists that specialize in its study hold conferences in some fabulous locations. In pursuit of the truth about this devastating pathogen I have been fortunate to have visited the South of France (where delegates to the rice blast meeting were enticed into a bull ring and attacked by a mild-mannered young bull) and the north coast of Japan. Let me tell you how this fungus works.

Magnaporthe is an ascomycete that forms those flask-shaped fruiting bodies called perithecia (Chapter 4). But this sexual phase rarely—or perhaps never—develops in an infected crop, and ascospores play no role in the disease at all. It is the asexual or conidium-producing segment of the life cycle that attacks rice plants.[4] Trouble begins for the rice plant when a pear-shaped conidium drifts onto its surface. Impact triggers the release of a sticky globule of mucilage from one end of the spore that binds the fungus to the waxy cuticle. If the leaf is wet with dew, the conidium germinates, and a germ tube travels a short distance from the spore before adhering at its tip and swelling into a bulbous structure called the appressorium (Figure 9.5). This is an infection apparatus and is one of the more remarkable structures formed by fungi.[5] Within a few hours the appressorium becomes blackened with melanin, fills itself with glycerol, becomes highly pressurized, and plunges straight through the leaf surface. In mechanical terms it ranks as the most powerful pathogen on the planet: pressure inside an appressorium can exceed 80 atmospheres. I have mentioned earlier that appressoria can puncture bulletproof vest material. This was discovered by Richard Howard, a scientist at the DuPont Company, who is responsible for much of the groundbreaking research on appressorial function. A "back-of-the-envelope" calculation shows that the pressure exerted against the rice leaf is considerably greater than the pressure exercised by the toe of a 400-pound dancer

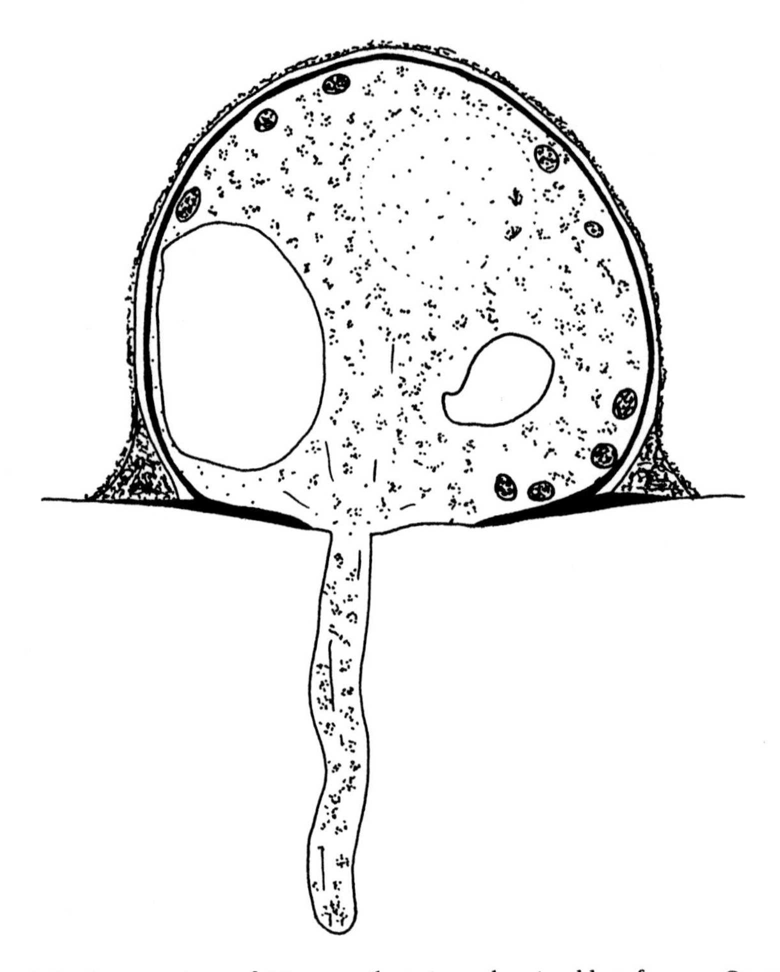

Fig. 9.5 Appressorium of *Magnaporthe grisea*, the rice blast fungus. Once the domed cell becomes highly pressurized, a filamentous penetration hypha extends from the base of the appressorium and pushes through the underlying surface. Drawing traced from electron microscopic image published by R. J. Howard, T. M. Bourett, & M. A. Ferrari. In: K. Mendgen & D.-E. Lesemann, eds., *Electron Microscopy of Plant Pathogens* (Berlin & Heidelberg: Springer-Verlag, 1991), 251–264.

engaged in a brief pirouette. The geneticist John Hamer once compared the force exerted by the fungus to the effect of a lighter woman wearing stiletto heels walking on someone's back. I would not be surprised to learn that John has tested his prediction by experiment. Metaphors aside, the mechanism of appressorium action is a wondrous scientific story.

Once the appressorium has stuck itself to the leaf by secreting an "O-ring" of glue, it deposits a veneer of pigment molecules against the inner surface of its cell wall. Within a couple of hours, a distinct melanin-rich layer has been added to the wall. If the formation of this layer is inhibited using a chemical fungicide that targets the biochemical pathway that the fungus uses to manufacture melanin, the appressorium will never penetrate the leaf. Similarly, the transparent appressoria of albino mutants of *Magnaporthe* are ineffective at plant penetration. Clearly, the change in color is linked to the mechanism of invasion. Most cell walls are quite leaky structures that allow small molecules to pass freely in either direction, but a melanized wall turns the cell into an impregnable container. This is critical for the rice blast fungus because it must accumulate and hold on to tremendously high concentrations of glycerol and other compounds that allow it to become pressurized. The mechanism of pressurization is osmotic: water diffuses into the cytoplasm in response to the rising glycerol content, and the melanin layer traps glycerol from diffusing out of the cell. Pressure rises and as the base of the appressorium begins to yield, a slender hypha is forced into the underlying leaf (Figure 9.5).

The pressure inside the cell has been measured using a number of different methods, and the force exerted during plant penetration has been determined by a group of German researchers whose work was published in the journal *Science*.[6] Force measurements from single hyphae present a considerable technical challenge. The German group solved the problem by allowing appressoria to indent a metal sandwich filled with breast-implant gel. They used a laser to measure the precise depth of the indentation, and from this they determined the force that must have been exerted by the appressorium from the force needed to indent the gel. An ingenious experiment indeed, although one can imagine the response to an article in a newspaper that explained how hundreds of thousands of German marks were spent to enable a group of researchers to watch fungi squeezing slivers of breast implant gel. For my own part, I've spent hundreds of thousands of taxpayers' dollars measuring how much force other kinds of hyphae exert, although I used Norwegian strain gauges that have no value in cosmetic surgery.

The rice blast fungus has been studied intensively by geneticists for many years, and its hereditary riches are now being mined by researchers who are sequencing the pathogen's entire genome. *Magnaporthe* has seven chromosomes that contain about 3 times more DNA than the yeast *Saccharomyces* and 80 times less than a human. The relationship between the amount of DNA in each nucleus and the complexity of an organism is quite messy. Many plants and amphibians, for example, have much more DNA than humans. So if we are going to argue that humans, rather than newts, are the peerless representatives of God's creative talent, we must also concede that we may fall short of being 80 times more impressive than the rice blast fungus. Yeast has about 6,000 genes that encode functional proteins, and humans have 30,000 to 40,000, depending on whom one listens to; the information content of the rice blast fungus presumably sits somewhere between these extremes. Less than a human, more than a yeast.

The reason so much effort is being expended on sequencing the rice blast genome is that there is a high probability of finding distinctive *Magnaporthe* genes whose products have not been found in any other organisms (yet). Because *Magnaporthe* has been engaged as a cereal predator for millions of years, we can anticipate genes that orchestrate the intimate relationship between the fungus and its hosts: genes that control the timing of conidium germination and melanin synthesis, genes that are turned on only when the fungus penetrates the epidermis of the plant, others expressed when it begins to feed on the cytoplasm, and so on. Yeast has none of these. Another pathogen of plants whose genome is under careful scrutiny is the cause of potato blight, *Phytophthora (fy-top-thora) infestans*. Its helices of DNA accommodate 250 billion of the coding letters (A, T, G, and C), called nucleotides, which specify the amino acid building blocks of proteins. "Blight" has a sixfold larger genome than "blast," but again, the actual number of functional genes is unknown.

Phytophthora is the most infamous fungus, linked forever with the Irish famine of the 1840s and the resulting diaspora that has placed a bar with an Irish name (few of these qualify as "Irish pubs") in most towns in the United States. The British are often blamed for the catastrophe. Millions of people were never far from starvation even in blight-free years, and the fact that this situation was maintained, at least in part, by British landowners is an appalling episode in my homeland's history. But to place blame wholly upon the British ignores the complexity of the tragedy. The introduction of the potato to Ireland allowed the population to swell to 8 million by the 1840s, the majority of whom (6 million perhaps) were utterly dependent on bumper crops of potatoes for their own sustenance and as the commodity used to meet rent payments. The Irish were poised for annihilation. And then we must consider the pathogen and its predilection for unusually wet climates. As the temperature plunged late in the summer of 1845, and weeks of fog and continuous rain ensued, *Phytophthora* got the better of the crop. A million people starved to death and a million or more emigrated across the Atlantic. I quote from a single article in the *Kilkenny Journal* (December 4, 1846), concerning an inquest held by a coroner's office following the discovery of the bodies of a 30-year-old mother and her three children:

> The bodies were brought to a house on the road side, the nearest that could be procured, by the police—they presented a truly heart-rending spectacle, partially covered with filthy rags saturated with mud, and frozen, having been exposed to the inclemency of the weather. The hand of one child, and part of the foot of another, had been devoured by rats. Doctor Gwydir, of Freshford, made a minute post mortem examination of the bodies of the mother and eldest daughter, a child about 9 years old. The Doctor was unable to detect in the stomach or the bowels of the mother a trace of food having entered for more than twenty hours before death. The child's stomach contained a very small quantity of half-digested potatoes.

Within a few years, the population had been halved by continuing malnutrition and emigration. The effects of the famine are evident today in the fact that Ireland's population remains below 4 million.

Potato blight has not disappeared. *Phytophthora infestans* causes $3 billion in losses every year, qualifying the pathogen as the most devastating living enemy of global agriculture. Most plant infections are caused by a single strain of the pathogen that was thought to have been the same one that precipitated the Irish famine. By extracting DNA from herbarium specimens of potato leaves collected at the time of the famine, researchers have found that this is not the case; some other strain destroyed potatoes in the 1840s. Plant breeders searching for resistant strains of potato are interested in locating the pathogen's birthplace, reasoning that wild potatoes that evolved alongside *Phytophthora* must have developed strong defenses. Areas of extraordinary genetic diversity are particularly fruitful hunting grounds, and a Mexican origin for *Phytophthora* seems logical because the greatest variety of strains coexist in the Toluca Valley north of Mexico City. But some researchers are convinced that it sprang from somewhere else, probably in South America, which also happens to be the ancestral home of the potato.

Phytophthora is another oomycete fungus, related to the microorganism that causes human pythiosis, and like other water molds it produces swimming zoospores. These develop in sporangia that can be dispersed by wind. If a sporangium drifts onto a wet leaf, the spores will be expelled into the film of water and charge over the waxy cuticle before forming a cyst. Subsequent development is similar to the rice blast fungus: a germ tube grows from the cyst, forms an appressorium, and pierces the epidermis with a penetration hypha. Early in its infection, *Phytophthora* makes haustoria and feeds from the potato plant without widespread destruction of its tissues. But after a few days, tissue damage is profound and the pathogen feeds on the increasingly soupy remains of its dying host. Just five days after entering the leaf, *Phytophthora* reemerges through open stomata and releases its airborne sporangia from the tips of aerial hyphae. As many as 300,000 sporangia can be launched each day from a single blotch on a potato leaf. Washed from the leaves, the sporangia release their spores into the soil to infect the buried tubers. Potatoes that appear healthy at the time of harvest can rot in storage, a process accelerated by bacteria that melt the starchy tissue into a slimy mess.

Because the fungus does not usually spread directly from the leaves down to the tubers, part of the crop can be salvaged by removing the infected leaves and stems when the blotches appear. A Belgian scientist discovered this preventive measure in 1845, but nobody in Ireland was aware of it. This was a time when potato blight was blamed on the recent introduction of the steam engine, on hidden volcanoes, and on the devil. There is irony then that an English clergyman, the Reverend Miles Berkeley, was responsible for identifying *Phytophthora* as the cause of the disease symptoms.[7]

More than fifty species of *Phytophthora* are recognized, and besides potatoes, they infect almost every kind of broad-leaved plant, and cause tens of billions of dollars in damage to crops every year. A newly identified species is responsible for a disease called sudden oak death in California. Infected tanoaks, coastal live oaks, and black oaks have been reported along a 200-mile-long stretch of the northern Californian coastline from Monterey to Mendocino County, some of the prettiest countryside in the United States. The mycelium spreads underneath the bark and up into aerial tissues, and red ooze laced with infectious sporangia seeps from infected trees. Once the rot is established, ascomycete fungi and wood-boring beetles take hold and the oaks begin to disintegrate.

In the closing pages of this book, I'd like to add a note about mycoparasites, fungi that infect other fungi. *Mr. Bloomfield's Orchard* furnishes ample evidence for the inextricable links among fungi and death and decay, and it should not surprise you that fungi "have little fleas upon their backs" that "bite them,"[8] nor that some of these pests are their own relatives. The most conspicuous mycoparasites form fruiting bodies on fruiting bodies. The mycelium of *Asterophora* probes the blackening tissues of old milk caps and russulas and buds its own powdery mushrooms at the surface. Similarly, while most species of *Cordyceps* are encountered as insect pathogens, some infect the underground fruiting bodies of truffles. Farmed mushrooms are also at risk. The humid conditions needed for mushroom cultivation encourage conidial fungi that cause dry and wet bubble, shaggy stipe, and cobweb disease, whose symptoms include bubbling of dead tissue, malformation of the cap and stem, envelopment in parasitic mycelium, exudation of colored drops of fluid, and the emission of foul smells. In this condition, mushrooms do not sell well. Control is a problem because there are few chemical agents that poison a fungal parasite without damaging the fungal host. The use of sterilized casings (the damp soil that is layered over the mushroom beds to provoke fruiting) and careful control of the temperature and relative humidity in the growing rooms reduce disease outbreaks. The deliberate introduction of a third fungus, called *Trichoderma*, is also a successful approach. It seems that this soil microorganism will attack the parasites without harming the crop. The efficacy of *Trichoderma* may be due partly to its fantastic rate of growth: the mycelium expands so swiftly that it starves other fungi in its neighborhood. The aggressive behavior of *Trichoderma* is also utilized in efforts to control certain plant diseases.

The fact that some fungi eat other fungi is symptomatic of the behavior of all of the microorganisms discussed in this book. These diverse species are united by four common features. They penetrate their foods with invasive hyphae, feed by absorbing nutrients, reproduce by spore formation, and show a marked propensity for attacking other organisms. Obvious parasites are encountered among all fungal groups, but many of the apparently innocuous species will attack living organisms if opportunities present themselves.

That's all I have to say about fungi in this book, but the title deserves further explanation. My first glimpse of the mysterious world of mushrooms and molds was offered by Mr. Bloomfield's orchard—in the village of Benson in Oxfordshire—but my association with the orchard goes deeper. In my teens an elderly man moved into the cul-de-sac of twelve homes that was our neighborhood. My mother told me that he was a retired professor from Birbeck College in London, but she didn't know his area of expertise. His wife Nora and he were in their 70s then. While I was engaged in the sundry responsibilities of a young teenager—stealing beer, shoplifting, and committing miscellaneous acts of vandalism—the professor walked the footpaths around the village deep in contemplation and otherwise proceeded with his retirement. I noticed him often.

Against earlier indications to the contrary, I went to college at Bristol University to study biology. My favorite teacher was Mike Madelin, a mycologist who had appeared in a British television documentary called *The Rotten World About Us*, a week or so before I left home. He was a star among a group of quite brilliant botany lecturers. His simple, understated style suited the material he taught, and I hung on every word he uttered. Madelin made me want to study fungi, and I begged to work in his lab. Nick Read, the cell biologist who tested the limits of spore durability in Chapter 4, was similarly awestruck by Madelin and credits him with his own career choice. This kind of conversion (at 18 years old in my case), is more often associated with the work of an evangelist.

One morning in Bristol, Mike showed me a book called *The Nature of Toadstools*. It was written by C. T. Ingold, the man who discovered star-shaped spores in a babbling brook. I flipped through the first pages as we talked and at the end of the preface read "Benson, Oxfordshire, 1978, C.T.I." My face flushed. I knew the author. The old man, the retiree, my neighbor, was Ingold. I had decided to become a mycologist, and a year later discovered that I had grown up next to England's best-known fungal biologist. On my next trip to see my parents I visited Terence Ingold and we became friends. We walked through the countryside around Benson, often in the direction of a pub called The Three Bells in a village called Rokemarsh that served the incomparable Brakespear ale. (Writing this, I am again sitting in my shed in Ohio, 4,000 miles from the source of this elixir; it's 2:00 P.M. in Ohio, early evening in the pub.) Terence showed me fungi in the field and at home in his culture dishes. He spoke with delight about each species, introducing them as old friends. He also introduced me to Buller's work, which he revered.

During his career, Ingold had written a series of books on fungi. These had introduced many biologists to the fungi and influenced a whole generation of mycologists. In retirement, Ingold continued his researches in a small study in his house, working with a microscope and tools for transferring cultures, and agar media and Petri dishes sent by a former colleague in London. He published frequently, describing novel developmental pathways taken by fungi grown with a limited supply of nutrients, and named a new fungus in honor of his adopted home. This yeast is called *Bensingtonia ciliata*. He also visited Webster and collaborated with him on the solution to the ballistospore discharge mechanism.

In many ways, during my own career, I have been fascinated by the kinds of questions that interested Ingold. In particular, we have both been captivated by the circus of bizarre mechanisms unique to the fungi. Although I've always wanted to explore how living things work, I do not know how much my choice of research projects has been influenced by my predecessor. Nevertheless, I do find the irony of adopting such an unusual career path and then discovering one's proximity to one of its greatest practitioners superbly strange.

I saw Ingold last a few years ago. He was well into his 90s, but still walked, read voraciously, and maintained a firm grip on current research. After his beloved wife Nora died, he sold their home and moved to a small apartment in a group of houses built a few hundred yards away.

This retirement community was named Chiltern Close, after the chalk hills that surround this part of the Thames River Valley. The houses were built on land exposed when bulldozers purged the remnants of Mr. Bloomfield's orchard. Fungi are already working on the apartment rafters, food in the residents' kitchens, and even on the residents themselves. They are everywhere and will outlive us by an eternity.

Notes

Chapter 1. Offensive Phalli and Frigid Caps

1. Basidiomycota is the name of one of the major taxonomic groupings of fungi called phyla. When referring to the formal name of group, it is capitalized; otherwise, we refer to these organisms as basidiomycete fungi, or basidiomycetes. The same terminology is applied to all of the groups of fungi discussed in this book.

2. In 1926, members of a French sect became convinced that these obscene fungi arose from bird droppings and afflicted those who inhaled their odors with "horrible diseases." To deal with this problem, they attacked a parish priest who was possessed by the devil, cast spells on sect members, and (most incriminating of all) had sent the birds. The mythology of stinkhorns and other fungi is discussed in W. P. K. Findlay, *Fungi, Folklore, Fiction, & Fact* (Eureka, Calif.: Mad River Press, 1982).

3. E. Schaechter, *In the Company of Mushrooms. A Biologist's Tale* (Cambridge, Mass.: Harvard University Press, 1997).

4. G. W. Gooday & J. Zerning, *The Mycologist* 11, 184–186 (1997).

5. The flower of *Amorphophallus* is actually a group of flowers, or an inflorescence. Its stench is emitted by a structure called the spadix, which elongates from the center of the inflorescence and can reach a height of 3 meters. Putrescine and cadaverine are among the culprits for the smell, which is said to have caused some people to pass out. The same chemicals are released during the decomposition of proteins in an animal carcass. *Rafflesia arnoldii*, the stinking corpse lily, grows in Indonesia and Malaysia and produces the world's largest single flower (up to one meter in diameter). These are the best-known carrion flowers, but many other plants utilize the same pollination strategy.

6. In *Climbing Mount Improbable* (New York: W. W. Norton, 1996), 274, Richard Dawkins wrote: "The wings of bees can truly be called flowers' wings, for they carry flower genes just as surely as they carry bee genes."

7. High-speed digital cameras now available in some research laboratories can capture an astonishing number of still images in a single second. These cameras may enable someone to photograph the moment when the spore begins its flight.

8. The story of Webster's discovery is recounted in my essay that appeared in *Mycologia* 90, 547–558 (1998).

9. A. H. R. Buller, *Researches on Fungi*, vols. 1–6 (London: Longmans, Green, 1909–1934); reprinted by Hafner Publishing Company in New York in 1958. *Researches on Fungi*, vol. 7, published posthumously by Toronto University Press (1950).

10. T. D. Bruns et al., *Nature* 339, 140–142 (1989).

Chapter 2. Insidious Killers

1. John Aubry (1626–1697) commented upon the similarity between ringworm patterns on skin and the growth of fairy rings in his *Natural History of Wiltshire* (written between 1656 and 1691), which was edited by John Britton and published 150 years after Aubry's death (London: Wiltshire Topographical Society, 1847). Fairy rings are discussed in Chapter 3.

2. A. Casadevall & J. R. Perfect, *Cryptococcus neoformans* (Washington, D.C.: ASM Press, 1998).

3. A. T. Spear, *Hawaiian Sugar Planters Association Experimental Station Pathology and Physiology Series Bulletin* 12, 7–62 (1912).

4. B. L. Wickes et al., *Proceedings of the National Academy of Sciences USA* 93, 7327–7331 (1996).

5. Correction. An untreatable mucormycosis is exactly the kind of misfortune one would wish on one's worst enemy.

6. P. K. C. Austwick & J. W. Copland, *Nature* 250, 84 (1974), is a classic citation, but fascinating papers on all aspects of the equine disease appeared in other journals.

7. Canine pythiosis is probably quite common, but is mistakenly diagnosed as a neoplastic condition (caused by a tumor).

8. Only three cases of human pythiosis in the United States have been described in the medical literature since 1989 (and no previous cases were reported). This sample size is so small that even though two of the patients were Texans we have no idea if the Lone Star State is a hot spot for the pathogen. The third case was reported from Tennessee.

Chapter 3. What Lies Beneath

1. The filamentous form and invasive behavior of hyphae is reproduced in root hairs, pollen tubes, and other tip-growing plant cells, but these cells do not create branched structures that approach the complexity of mycelia.

2. H. N. Smith, J. N. Bruhn, & J. B. Anderson, *Nature* 356, 428–431 (1992).

3. For most purposes, the terms parasite and pathogen are interchangeable when referring to disease-causing fungi.

4. A few fungi, including the rusts, grow in the tissues of particular host plants and nowhere else. Species with this specialized feeding behavior are called obligate parasites or biotrophs. With few exceptions, these biotrophs cannot be cultured in the laboratory on agar medium.

5. According to the researchers who listen to the electrical activity of hyphae, fungi "laugh themselves senseless" when they read this mission statement.

6. The researchers did find, however, that the mice took much longer to die when infected with the protease-deficient strains of *Candida*; B. Hube et al., *Infection and Immunity* 65, 3529–3538 (1997); D. Sanglard et al., *Infection and Immunity* 65, 3539–3546 (1997).

7. The experiments are summarized in N. P. Money, *Fungal Genetics and Biology* 21, 173–187 (1997).

8. The phrase is appropriated from S. Jones, *Almost Like a Whale. The Origin of Species Updated* (London: Doubleday, 1999), xviii.

9. Secreted proteinases play the same tissue-degrading role when invasive cancer cells penetrate solid tissues and also during the implantation of the human embryo in the wall of the uterus.

10. Hyphal morphogenesis is explored by Frank Harold in *The Way of the Cell: Molecules, Organisms, and the Order of Life* (Oxford & New York: Oxford University Press, 2001).

Chaper 4. Metamorphosis

1. L. Weschler, *Mr. Wilson's Cabinet of Wonder* (New York: Pantheon Books, 1995).

2. Ascomycetes and insects both appeared about 400 million years ago and it is likely that some of them have been fighting ever since.

3. The quote comes from a plaque in the church of Wattisham in England and is discussed by Adrian Morgan (1995) in his beautifully illustrated book *Toads and Toadstools. The Natural History, Folklore, and Cultural Oddities of a Strange Association* (Berkeley, Calif.: Celestial Arts, 1995), 44.

4. The vasoconstricting properties of purified ergot toxins are employed to induce uterine contractions in childbirth and to stop bleeding, and to treat migraine headaches. The ergot fungus was used as an abortive agent by midwives in the sixteenth century.

5. Carpology is the study of fruits. The Tulasnes' *Carpologia* was concerned with fungal "fruits" (spore-producing structures) and "seeds" (spores), whether formed with a sexual partner or in solitude. The original was written in Latin and appeared between 1861 and 1865, but thankfully an English translation from 1931 is available in botanical libraries: L. R. Tulasne & C. Tulasne, *Selecta Fungorum Carpologia*, 3 vols., translated by W. B. Grove, edited by A. H. R. Buller & C. L. Shear (Oxford: Clarendon Press, 1931). The hand-colored illustrations in the Tulasnes' other masterpiece, *Fungi Hypogæi. Histoire et Monographie des Champignons Hypogés* (Paris: Apud Friedrich Klincksieck, 1851), are equally intoxicating. Only 100 copies of this book were printed. Reproductions of the Tulasnes' illustrations are always disappointing, and I hope that their absence in this chapter will stimulate some readers to seek out the original publications.

6. D. A. Sibley, *The Sibley Guide to Birds* (New York: Knopf, 2000).

7. It is difficult to comprehend why disciples of Aristotle's fundamentalist view of spontaneous generation overlooked the relevance of animal reproduction as an explanation for the proliferation of rodents, and that they continued to do so as late as the seventeenth century. (As evidence of human gullibility, however, we need look no further than the fact that most Americans remain convinced of the reality of the devil, angels, and extraterrestrial visitors.) By the nineteenth century, adherents of the theory of spontaneous generation had limited debate to the development of microorganisms.

8. A. de Bary, *Comparative Morphology and Biology of the Fungi Mycetozoa and Bacteria*, English translation by H. E. F. Garnsey, revised by I. B. Balfour (Oxford: Clarendon Press, 1887).

9. The story is recounted in G. C. Ainsworth's superb *Introduction to the History of Mycology* (Cambridge: Cambridge University Press, 1976). Among Ainsworth's most surprising findings was that some natural historians seriously disputed the living nature of fungi throughout the eighteenth century, and that as recently as 1804, the idea that fungi were the offspring of comets was not universally ridiculed.

10. P. M. Kirk et al. *Ainsworth and Bisby's Dictionary of Fungi*, 9th ed. (Wallingford, United Kingdom: CAB International, 2001).

11. *Geopyxis cacabus* was discussed by Sir John Burnett in his *Fundamentals of Mycology*, 2nd ed. (London: Edward Arnold, 1976), 151, but the original description and location of the brute have been lost to science.

12. This number is based on estimates of spore production for other ascomycetes by C. T. Ingold, in *The Fungus Spore,* edited by M. F. Madelin (London: Butterworths, 1966), 113–132.

13. The estimate comes from A. H. R. Buller, *Researches on Fungi,* vol. 1 (London: Longmans, Green, 1909). Although 7 trillion is an impressive number of anything, it is disturbing to note that Bill Gates has almost as many pennies (his worth at the time of writing is $59 billion). But even though Bill has been a fortunate merchant, he's like the rest of us who will (in all likelihood) never gaze on *Geopyxis cacabus.*

14. N. D. Reid & K. M. Lord, *Experimental Mycology* 15, 132–139 (1991).

15. Resistance to dessication may have been a significant aspect of the evolution of subterranean basidiomycetes (false truffles) from boletes. This was discussed in Chapter 1.

16. Although fictitious, the changes in morphology described in my movie are supported by a number of clues. (i) Molecular genetic evidence favors a close relationship between truffles and cup fungi. (ii) The fruiting bodies of an ascomycete called *Genea* resemble apothecia but open only through a wide pore on the upper surface. Its asci have the elongated shape of those found in a cup fungus, rather than the spherical asci of truffles, but other structural characteristics indicate that *Genea* is closely related to truffles. (iii) In some truffle species, fluid-filled veins (Louis Tulasne called them "venes aquifères") surrounded by asci spread through the fruiting body and open via pores on the outer surface. This structural feature is consistent with the evolution of truffles by some defect in the process of fruiting body emergence and expansion that created an intensely corrugated cup from which the modern structure evolved.

Chapter 5. The Odd Couple

1. Established in 1877, the University of Manitoba was the first university in western Canada.

2. The spores fell at speeds of between 0.5 millimeters and 6.1 millimeters per second. Since the distance between the two boundary threads was set at 4.6 millimeters, the spores spanned this distance in 0.8 seconds to 9.2 seconds. The microscope optics inverted the viewed image, so that the spores appeared to move upward rather downward. A. H. R. Buller, *Researches on Fungi,* vol. 1 (London: Longmans, Green, 1909).

3. A. H. R. Buller, *Nature* 80, 186–187 (1909).

4. See note 9 in Chapter 1.

5. A Japanese entomologist has scooped me here. Nobuko Tuno of

Nagasaki University has found that after a single fly visits a bracket fungus called *Elfvingia applanata,* it can deposit hundreds of thousands of its basidiospores in its feces. N. Tuno, *Ecological Research* 14, 97–103 (1999).

6. I refer, of course, to the title of Daniel C. Dennett's book, *Darwin's Dangerous Idea. Evolution and the Meanings of Life* (New York: Simon & Schuster, 1995).

7. In a published description of a new species the author appends the term species nova to the Latin binomial (e.g., *Morchella mirabilis* sp. nov.). Subsequent references to this species replace sp. nov. with the author's name (*Morchella mirabilis* Money).

8. Lloyd explained to his readers that Mr. McGinty was the brother-in-law of Sairey Gamp, a most disagreeable midwife created by Charles Dickens in *Martin Chuzzlewit.* Curtis Lloyd's more serious contributions to the field of fungal taxonomy are assessed by W. H. Blackwell & M. J. Powell, *Mycotaxon* 58, 353–374 (1996).

9. C. S. Millard (photographs by G. Steinmetz), *National Geographic Magazine* 200, 110–125 (July 2001).

Chapter 6. Ingold's Jewels

1. Ingold recounts the story of his discovery in *A Century of Mycology,* edited by B. Sutton (Cambridge: Cambridge University Press, 1996), 39–52.

2. Protozoan is an informal term for single-celled eukaryotes, including amoebas and flagellates, classified within Kingdom Protista (the protists).

3. *The Guinness Book of Records* (New York: Facts On File, 1992). Unfortunately, Michel is not mentioned in more recent editions.

4. J. Webster, in *Evolutionary Biology of the Fungi,* edited by A. D. M. Rayner, C. M. Brasier, & D. Moore (Cambridge: Cambridge University Press, 1987), 191–201.

5. P. A. Cox, *The American Naturalist* 121, 9–31 (1983).

6. Mucilage at the tips of the appendages may mediate initial attachment, and the secretion of adhesive is probably initiated by a sensory mechanism that responds quickly to physical contact with a surface.

7. In the absence of incontrovertible data it is also possible that the pressure-driven mechanism of ascospore discharge evolved first in saltwater and later became widespread among terrestrial ascomycetes.

8. L. Berger, et al., *Proceedings of the National Academy of Sciences, U.S.A.* 95, 9031–9036 (1998).

9. J. M. Kiesecker, A. R. Blaustein, & L. K. Belden, *Nature* 410, 681–684 (2001).

10. W. C. Coker, *The Saprolegniaceae, With Notes On Other Water Molds* (Chapel Hill: University of North Carolina Press, 1923), 124–125.

11. N. P. Money, J. Webster, & R. Ennos, *Experimental Mycology* 12, 13–27 (1998).

12. The electron microscope analyses were limited to two species of *Achlya*, and giving Hartog the benefit of the doubt, functional flagella might be present in other species. Nevertheless, even in *Saprolegnia* spores are still shot from the sporangium under pressure and swim only after discharge into the water.

13. S. J. Gould, *Wonderful Life. The Burgess Shale and the Nature of History* (New York: Norton, 1989).

14. I. K. Ross, *Biology of the Fungi. Their Development, Regulation, and Associations* (New York: McGraw-Hill, 1979).

Chapter 7. Siren Songs

1. J. R. Raper, *American Journal of Botany* 26, 639–650 (1939).

2. D. S. Thomas & J. T. Mullins, *Science* 156, 84–85 (1967).

3. N. P. Money & T. W. Hill, *Mycologia* 89, 777–785 (1997).

4. Alma Wiffen Barksdale, at the New York Botanical Garden, was responsible for many pioneering studies on water mold reproduction. She postulated that hermaphrodite strains evolved from heterothallic strains: A. W. Barksdale, *American Journal of Botany* 47, 14–23 (1960).

5. Mycologists now regard the zygospore as a sexual spore formed inside a structure called the zygosporangium.

6. A. D. M. Rayner & L. Boddy, *Fungal Decomposition of Wood: Its Biology and Ecology* (New York: John Wiley & Sons, 1988).

7. R. B. Peabody, D. C. Peabody, & K. M. Sicard, *Fungal Genetics and Biology* 29, 72–80 (2000).

8. Buller and other mycologists suggested that the cystidia functioned in separating the gills. Their true function as trusses was elucidated by David Moore in the 1990s. Details are given in D. M. Moore, *Fungal Morphogenesis* (Cambridge: Cambridge University Press, 1998).

9. Research on asexual reproduction by Klebs was discussed in Chapter 6.

10. The harvest of wild mushrooms from the forests of the Pacific Northwest in 1992 was estimated at almost 2 million pounds with a value of $40 million. Global trade in fruiting bodies of *Tricholoma matsutake* (matsutake) is estimated

at $3 billion, and at $1.5 billion for *Cantharellus cibarius* (chanterelles). Data cited from *Fungal Conservation: Issues and Solutions*, edited by D. M. Moore, et al. (Cambridge: Cambridge University Press, 2001).

Chapter 8. Angels of Death

1. From a scientific point of view it is important to recognize that the mushroom is one facet of a life cycle that includes different stages of mycelial development (Chapter 1). *Amanita muscaria*, for example, is a species of basidiomycete fungus that produces a fruiting body whose common name is the fly agaric.

2. Exploiting the liver's powers of regeneration, surgeons treating patients poisoned by amatoxins have elected for partial transplants in which the left lobe of the damaged organ is removed and replaced temporarily with a donor lobe (along with the appropriate blood vessels and biliary drainage into the intestine). The donor lobe supports the necessary liver function for a few days until the original organ heals. Interestingly, other organs are damaged in serious poisonings, but the early presentation of liver failure masks more widespread injury to the body.

3. While some guidebooks refer to the distinctive flavors of mushrooms, one should *never* taste a wild mushroom without first obtaining an expert opinion. Obviously, I am not to be trusted on the finer points of mushroom identification, which is something my dinner guests must never learn.

4. Connie Barlow discusses the Osage orange in *The Ghosts of Evolution: Nonsensical Fruit, Missing Partners, and Other Ecological Anachronisms* (New York: Basic Books, 2000).

5. Jim Basinger, at the University of Saskatchewan, is a strong advocate of this hypothesis.

6. By "designed," I do not mean to imply, of course, that there is any master plan for the structure and composition of mushrooms. Today's mushrooms are effective structures for spore release because their ancestors were also competent at producing and dispersing offspring (spores). It seems most likely that a compound which kills insect grubs evolved as an insecticide, although in the absence of any evidence (this all happened a very long time ago), it is certainly possible that it was originally selected because it performed a completely different function that enhanced the operation of the fruiting body. Biologists make educated guesses about the initial functions performed by specific features of an organism, but we have been spectacularly wrong on occasion. Take

feathers, for example. Rather than evolving as structures crucial to the avian flight mechanism, there is excellent fossil evidence showing that they were elaborated first as an insulating material in flightless animals and were not co-opted until much later as wing coverings.

7. Dennis Benjamin recounts the story of the murder of Emperor Claudius I in his authoritative and entertaining book on mushroom poisoning, *Mushrooms: Poisons and Panaceas* (New York: W. H. Freeman, 1995). Fruiting bodies of *Amanita caesaria* were a favorite food for Claudius, an appetite exploited by his wife Agrippina, who murdered him by adding the lethal juice of *Amanita phalloides* to his dinner.

8. M. C. Cooke (1825–1914) was a prolific author of books on fungi, algae, and plants, and, in "a foolish moment, perhaps," penned a guide to British reptiles. He discussed mushroom intoxication in *The Seven Sisters of Sleep: Popular History of the Seven Prevailing Narcotics of the World* (London: J. Blackwood, 1860).

9. According to some sources, Satan's boletes have the odor of rotting flesh, but my specimen was quite fragrant.

10. C. McIlvaine & R. K. Macadam, *One Thousand American Fungi. How to Select and Cook the Edible; How to Distinguish and Avoid the Poisonous*, rev. ed. (Indianapolis: Bowen-Merrill, 1900, 1902). Here are a few of McIlvaine's recommendations. "Cut in slices . . . it is a tender, agreeable food" (phallic mushroom eggs). "Intensely bitter when slightest tinge of yellow is visible . . . A little lemon juice or sherry improves it" (puffball). "The squamules (scales) must be cut away or the dish will be rough. With many this is a prime favorite" (a black, scaly bolete known as the old man of the woods, whose flesh turns red when bruised). "The flavor is higher than that of *C. comatus*. It should be cooked as soon as gathered" (*Coprinus atramentarius*—I assume he did not drink wine). He took "special pains" to ascertain the edibility of Satan's boletes, and found them "remarkably fine eating" (see previous note).

11. D. Arora, *Mushrooms Demystified: A Comprehensive Guide to the Fleshy Fungi*, 2nd ed. (Berkeley: Ten Speed Press, 1996).

12. The Darwin Award was established by author and Internet consultant Wendy Northcutt "to commemorate those who improve our gene pool by removing themselves from it." From *The Darwin Awards: Evolution in Action* (New York: Dutton, 2000).

13. In addition to the turkeys, 14,000 ducklings died and numerous flocks of game birds were poisoned following consumption of the same peanut meal, which had been imported from Brazil.

14. L. Ruess, *Nematologica* 41, 106–124 (1995).

15. The term mold is applied very broadly to any microscopic fungus. Only the basidiomycete and ascomycete fungi with macroscopic fruiting bodies are excluded from this catchall title.

16. One cubic meter is equivalent to 1,000 liters. Human lung capacity varies between 3 liters and 7 liters of air, so even at concentrations below 50 spores per cubic meter, one or more spores will enter the lungs with every few breaths. Even in such clean environments, long-term exposure to low concentrations of spores may be unhealthy for some individuals.

Chapter 9. Mr. Bloomfield's Orchard

1. H. C. Hoch, et al., *Science* 235, 1659–1662 (1992).

2. Drop formation in rusts was first mentioned in 1912 by Dietel, who was unaware of Buller's monumental research on spore discharge in mushrooms. Buller pointed to the similarity of the discharge process in rusts and mushrooms in 1915, and published the authoritative study on this topic in the third volume of his *Researches on Fungi* in 1924. The evolutionary affinity between rusts and mushrooms was first proposed by Anton de Bary in 1881, based on the resemblance between the basidia of jelly fungi and the spore-producing "promycelium" (= basidium) that develops from the rust teliospore. But much credit goes to Reginald for his meticulous description of the spore discharge mechanism and for realizing its phylogenetic significance. A process of this complexity, one of utter uniformity among all basidiomycetes, is likely to have arisen just once. Buller is also credited with the discovery of spore discharge associated with drop formation in smuts, indicating the kinship of this second group of plant pathogens with mushrooms and rusts.

3. B. A. Roy, *Nature* 362, 56–58 (1993).

4. The asexual phase (or anamorph) was named *Pyricularia oryzae*, but if you remember the business about naming ascomycetes from Chapter 4, you'll recognize that the name for the sexual phase (the teleomorph), *Magnaporthe grisea*, takes precedence over the anamorph's name.

5. There is a vast literature on rice blast appressoria. A lavishly illustrated book chapter by Adrienne Hardham, which examines the cell biology of plant infection and surveys recent work on appressorial function, appears in *The Mycota*, vol. 8, *Biology of the Fungal Cell*, edited by R. J. Howard & N. A. R. Gow (Berlin & Heidelberg: Springer-Verlag, 2001), 91–123.

6. C. Bechinger, et al., *Science* 285, 1896–1899 (1999), and commentary article by N. P. Money, *Nature* 401, 332–333 (1999).

7. The remarkable story of Berkeley's work has been described by many authors. David Moore provides a lively account in *Slayers, Saviors, Servants and Sex* (New York: Springer-Verlag, 2001).

8. I quote from the well-known proverb: "Big fleas have little fleas upon their backs to bite them, and little fleas have lesser fleas, and so ad infinitum." A variant upon this saying first appeared in print in Jonathan Swift's *On Poetry, A Rhapsody* (Dublin & London, 1733).

Index

Printed in the United States
57740LVS00004B/142-144

9 780195 171587